D0073486

Silicon Geochemistry
and
Biogeochemistry

Silicon Geochemistry
and
Biogeochemistry

Edited by

S. R. Aston
Department of Environmental Sciences,
University of Lancaster,
Lancaster, UK

1983

ACADEMIC PRESS

A Subsidiary of Harcourt Brace Jovanovich, Publishers
London · New York
Paris · San Diego · San Francisco · São Paulo
Sydney · Tokyo · Toronto

ACADEMIC PRESS INC. (LONDON) LTD.
24/28 Oval Road
London NW1

United States Edition published by
ACADEMIC PRESS INC.
111 Fifth Avenue
New York, New York 10003

British Library Cataloguing in Publication Data

Silicon geochemistry and biogeochemistry.
 1. Silicon compounds
 I. Aston, S. R.
 547'.08   QD412.S6

ISBN 0-12-065620-5

Filmset in Great Britain by Eta Services (Typesetters) Ltd., Beccles, Suffolk
and printed by Galliard Printers Ltd., Great Yarmouth, Norfolk

___Contributors___

S. R. Aston*, Department of Environmental Sciences, University of Lancaster, Lancaster, LA1 4YQ, UK

S. E. Calvert, Department of Oceanography, University of British Columbia, Vancouver, British Columbia, V6T 1W5, Canada.

D. C. Hurd, Shell Development Corporation, Geological Engineering Division, P. O. Box 48, Houston, Texas 77001, USA

R. Macdonald, Department of Environmental Sciences, University of Lancaster, Lancaster, LA1 4YQ, UK

F. T. Mackenzie, Department of Geological Sciences, Northwestern University, Evanston, Illinois 60201, USA

C. P. Spencer, Marine Science Laboratories, University College of North Wales, Menai Bridge, Anglesey, LL59 5EH, UK

R. Wollast, Laboratoire d'Océanographie, Université libre de Bruxelles, Av. F. D. Roosevelt 50, 1050 Bruxelles, Belgium.

* Present address: International Laboratory of Marine Radioactivity, Musée Océanographique, MC98000, Monaco.

v

Preface

Silicon is an element of major interest in geochemistry and in the biochemistry of certain organisms. The lithosphere contains some 27% of silicon by weight, and most aspects of geochemistry directly involve the behaviour of silicon-containing minerals or their weathering products. As an essential nutrient, and the basis of the skeletal material of certain organisms, silicon occupies a dominant position in biogeochemistry. It is because of these important roles that silicon should be the subject of a book devoted directly to its geochemistry and biogeochemistry. This book is an attempt to draw together some of the more important aspects of this topic using a series of reviews.

Emphasis is placed on silicon in the Earth's surface environment, with the marine behaviour of silicon taking a dominant position. However, the book begins with a discussion of silicon in igneous and metamorphic rocks. Silicon in the mantle, transfer from the mantle to the crust, and silicon in the continental crust are considered. This is followed by a treatment of the global cycle of silicon in the Earth's surface environment. Weathering, transport on the global scale, sinks, and mass balances are considered. The third chapter examines the chemical nature of silicon in natural waters and the atmosphere, paying particular attention to the estuarine environment. The following contribution deals with the marine biogeochemistry of silicon, stressing its role as an essential nutrient. The final two chapters are discussions on the sedimentary geochemistry of silicon, especially in modern sediments, and the physical and chemical properties of siliceous skeletons.

Each contribution is intended to be self-contained for the most part, and this has led to some areas of overlap, but they are necessary to allow discussion of certain topics within the framework of the individual chapters. I hope that the drawing together of the geochemistry and biogeochemistry of silicon into this book will be an aid to students of geochemistry, geology, aquatic biology, and ecology. Also, if it encourages research into this subject it will have obtained one of its most important goals.

Monaco, 1982   _S. R. Aston_

Contents

1

Silicon in Igneous and Metamorphic Rocks

R. Macdonald

Department of Environmental Sciences
University of Lancaster, UK

1.1 SILICON IN THE EARTH'S MANTLE

1.1.1 Introduction

All geological processes involved in crust–mantle interactions have tended to concentrate silicon (Si), relative to its mantle proportions, into the upper parts of the continental crust. In this chapter, the nature of the major processes effecting Si transfer from mantle to crust is discussed, and the compositions and proportions of the dominant silicate materials tabulated.

1.1.2 Composition and Mineralogy of the Mantle

Seismic evidence indicates that most magmas are generated within the mantle at depths between 50 and 170 km, though in certain regions the sources may be as deep as 400 km (Yoder, 1976). This latter depth corresponds approximately to the top of the transition zone, such that magma generation seems to be restricted to the upper mantle.

In all silicates within this depth range, Si is coordinated to 4 oxygens (^{IV}Si). Liebaü (1972) gives a full account of the structural classification of silicates with tetrahedral coordination and this need not be repeated here. $^{IV}Si \rightarrow {}^{VI}Si$ transformations are likely to become important only in the lower two-thirds of the transition zone (500–900 km). Hazen and Finger (1978) listed available

1

data on the $^{IV}Si \rightarrow {}^{VI}Si$ transformations in common upper mantle and crustal silicates.

The bulk of geophysical, geochemical, and petrological data supports a peridotitic upper mantle. Study of the xenoliths carried up in diatremes and in basaltic lavas suggests that the upper mantle is not homogeneous, mineralogically or chemically. A wide range of mantle-derived rock types is found in kimberlites, for example (Harte, 1978): dunite (olivine), harzburgite (olivine + orthopyroxene ± garnet ± spinel), lherzolites (olivine + orthopyroxene + clinopyroxene ± garnet ± spinel ± mica), garnet pyroxenites, various types of eclogite (garnet + pyroxene) and amphibole-bearing and mica-rich rocks. Several modes of origin are applicable to these rocks:

(i) They are source mantle, either fertile or partially barren. Such mantle may represent "rejuvenated" barren mantle.

(ii) They are barren mantle, having had a basalt fraction removed.

(iii) They represent cognate xenoliths, precipitated from a range of mafic to ultramafic liquids.

(iv) They represent any of cases (i) to (iii) but re-equilibrated to new mineral assemblages under varying P–T conditions.

(v) They represent mafic liquids trapped within the mantle and recrystallized to appropriately high-pressure assemblages.

Although there is considerable controversy as to which mode, or modes, of origin may apply to each xenolith type, the consensus of opinion is that the most likely contenders for primary mantle status are garnet lherzolite nodules from kimberlites. Further debate, however, revolves round whether it is the coarse (granular) or sheared textural varieties of nodules which are the more pristine (Kushiro, 1973; O'Hara et al., 1975; Boyd and Nixon, 1975, 1978; Meyer, 1977; Carswell, 1980).

Such complexities of interpretation cannot be explored here and chemical data for rocks and minerals of both varieties of nodule are employed to discuss mantle chemistry. Table 1.1 compares estimates of the major element composition of fertile (undepleted) peridotite based mainly on garnet lherzolite xenoliths. Assuming that they give a reasonably accurate guide to mantle chemistry, it appears that the relative variation in Si content is around 10%. The spreads of ratios of Si to the other major components vary with the abundance of those components. Thus Si/Mg shows about a 10% spread, and Si/Na a 300% spread. Abundances of such elements as Ca, Na, K, Al, and Fe are, however, very susceptible to small amounts of fusion and to contamination of the nodules by the host magma. It is not certain therefore how much of the compositional spread reflects heterogeneity in truly pristine mantle (Carswell, 1980).

TABLE 1.1   Various estimates of the major element composition of "fertile" upper mantle and average compositions of garnet lherzolite nodules from kimberlites (recalculated to 100% volatile-free and with all Fe expressed as FeO).

Element (weight per cent)	I	II	III	IV	V	VI	VII
O	44.0	44.6	43.9	44.7	44.8	43.9	43.8 –44.2
Mg	24.9	26.9	25.9	25.6	25.7	26.8	24.6 –24.6
Si	20.7	21.7	20.5	22.2	21.8	20.5	20.9 –21.5
Fe	6.5	5.0	6.7	5.0	5.1	6.2	6.7 – 7.0
Ca	1.7	0.61	1.3	0.80	0.81	0.95	1.7 – 1.2
Al	1.4	0.43	0.98	0.86	1.0	0.92	1.4 – 1.2
Cr	0.21	0.19	0.15	0.28	0.21	0.29	0.16– 0.14
Na	0.19	0.07	0.16	0.10	0.13	0.18	0.07– 0.07
Ni	0.16	0.23	0.18	0.26	0.23	—	0.16– 0.13
Mn	0.12	0.09	0.09	0.09	0.08	0.09	0.09– 0.06
Ti	0.06	0.04	0.10	0.04	0.05	0.10	0.07– 0.05
K	0.01	0.12	0.02	0.09	0.12	0.12	0.02– 0.02
P	—	0.03	0.00	0.01	0.02	—	trace
Si/Mg	0.83	0.81	0.79	0.87	0.85	0.76	0.85– 0.87
Si/Fe	3.2	4.3	3.1	4.4	4.3	3.3	3.1 – 3.1
Si/Ca	12.2	35.6	15.8	27.8	26.9	21.6	12.2 –17.7
Si/Al	14.8	50.5	20.9	25.8	21.8	22.3	14.6 –18.5
Si/Na	109	310	128	222	168	114	299– 307
Si/K	2070	181	1025	247	182	171	997–1022

  I. Estimated upper mantle composition (Harris *et al.*, 1967).
 II. 16 coarse (granular) garnet lherzolite nodules in Lesotho kimberlites (Carswell, 1980, Table 5, col. C).
III. 6 porphyroclastic (sheared) garnet lherzolite nodules in Lesotho kimberlites (Carswell, 1980, Table 5, col. D).
 IV. 5 common peridotite nodules from Matsoku kimberlite, Lesotho (Carswell, 1980, Table 5, col. E).
  V. 23 garnet lherzolite nodules in kimberlites of Kimberley area, South Africa (Carswell, 1980, Table 5, col. A).
 VI. 6 garnet lherzolite nodules in kimberlite of Udachnaya Pipe, Yakutia, USSR (Carswell, 1980, Table 5, col. G).
VII. Range of estimated mantle compositions, based on the chemistry of ultramafic lavas (Bickle *et al.*, 1976, Table 1).

The estimated mantle compositions will serve, though, as a basis to assess Si behaviour during mantle fusion (Section 1.2.1.2).

Whatever its composition, or range or compositions, fertile mantle rock may exist as three distinct mineral assemblages, each indicative of different P–T conditions of equilibration (Fig. 1.1). In terms of increasing pressure, the assemblages are:

olivine + Al-poor orthopyroxene + Al-poor clinopyroxene + plagioclase
olivine + aluminous orthopyroxene + aluminous clinopyroxene ± spinel
olivine + Al-poor orthopyroxene + Al-poor clinopyroxene +

pyrope-rich garnet.

Plagioclase peridotite is likely to exist only in oceanic regions or in continental areas where attenuated crust overlies areas of high heat flow. Most basalts are generated within the stability fields of spinel- and garnet-peridotites; Yoder (1976) argued that, in fact, garnet peridotite was the normal parental material.

Various estimates of model mantle mineralogy are provided in Table 1.2. Some of the variation reflects variation in the pristine upper mantle; some must be due to the partly residual nature of the specimens employed.

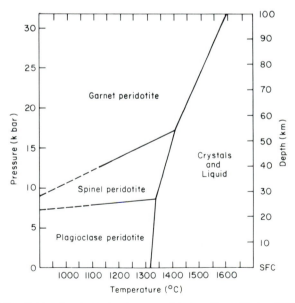

FIG. 1.1 Stability fields of various aluminous peridotites (from Yoder, 1976, Figs 2–10). SFC = surface.

TABLE 1.2   Estimates of mantle mineralogy, compiled by Wyllie (1971).

	Spinel Peridotite		Garnet Peridotite	
	Carter (1966)	Harris *et al.* (1967)	Harris *et al.* (1967)	Ringwood (1969)
Olivine	55 ± 10	65.3	67	57
Orthopyroxene	27 ± 5	21.8	12	17
Clinopyroxene	14.5 ± 3.5	11.3	11	12
Spinel	3 ± 1	1.5	—	—
Garnet	—	—	10	14
Total	99.5	99.9	100	100

See Wyllie (1971) for above references to source papers.

1.1.3   Mineral Chemistry

Table 1.3 lists the compositions of olivines, pyroxenes, garnet, and spinel recorded in various lherzolitic nodules from African kimberlites. The data are presented as weight per cent of elements, in order to facilitate comparison with Table 1.1

The list is not an exhaustive summary of fertile mantle mineral chemistry. It ignores the possible presence in mantle rocks of such minor phases as mica, amphibole, and ilmenite and it assumes that nodules in kimberlite are truly representative of such mantle. However, the list, in conjunction with Table 1.2, serves to demonstrate that more than 80% wt. of the Si in the upper mantle resides in olivine and Ca-poor pyroxene.

1.2   PROCESSES TRANSFERRING SILICON FROM MANTLE TO CRUST

1.2.1   Magmatism

1.2.1.1   Introduction

The predominant process in transporting Si from the mantle into the crust is magmatism. Some 10 km^3 of magma is emplaced within the oceanic crust at constructive plate margins every year (Gass, 1978). Assuming an average density of 3000 kg m^{-3} and an average Si content of 23.7% wt. for this material, it may be inferred that ~8 × 10^9 t of Si are added to the oceanic crust each year. Even the deepest parts of this crust, which may never be exposed at the surface during the sea-floor spreading cycle, may well be subjected to extensive hydrothermal alteration by convective circulation of

TABLE 1.3 Compositions (in weight per cent) of major mineral phases in the upper mantle, based on lherzolite nodules.

	Olivine			Orthopyroxene			Clinopyroxene			Garnet			Spinel		
	1	2	3	1	2	3	1	2	3	1	2	3	3	4	5
O	43.90	43.72	44.38	46.92	46.23	46.63	47.47	45.26	44.43	47.00	44.97	44.34	36.93	32.09	34.04
Mg	30.83	29.79	31.02	20.87	20.29	21.90	9.95	12.56	10.08	12.67	12.99	11.82	8.16	7.70	8.40
Si	19.16	18.82	19.09	27.07	26.63	26.94	25.39	25.81	25.50	19.68	19.86	19.46	0.03	0.05	0.03
Fe	6.11	7.40	5.38	4.35	4.49	3.37	1.76	3.26	1.46	5.74	5.54	5.08	10.35	13.18	12.88
Ca	0.01	0.07	0.01	0.34	1.06	0.29	13.94	10.09	14.66	3.43	3.27	4.57	tr.	tr.	tr.
Al	tr.	0.04	0.02	0.45	0.67	0.48	1.49	1.17	1.23	11.48	10.63	10.10	7.22	4.79	8.20
Cr	—	0.04	0.01	—	0.20	0.25	—	0.48	1.22	—	2.07	4.27	37.01	41.58	35.79
Na	—	tr.	—	—	0.24	0.05	—	1.14	1.34	—	0.06	tr.	—	—	—
Mn	—	0.11	0.09	—	0.10	0.09	—	0.10	0.06	—	0.20	0.34	0.22	0.30	0.26
Ti	—	0.01	tr.	—	0.09	tr.	—	0.13	0.02	—	0.41	0.02	0.08	0.31	0.40

1. Averaged compositions from 8 garnet lherzolites, various African kimberlites (Bishop et al., 1978, nos. 740, 749, 1140, 1143 B, 1149, 1359, 1870/2).
2. Averaged compositions from 11 sheared lherzolite nodules, N. Lesotho kimberlites (Nixon and Boyd, 1973, Table 20B).
3. Averaged compositions from 16 granular lherzolite nodules, N. Lesotho kimberlites (Nixon and Boyd, 1973, Table 20A).
4. Averaged compositions of spinel in 3 spinel lherzolite nodules from Africa (Smith and Dawson, 1975, Table 1, specs. A, B, D).
5. Averaged compositions of spinel in 3 garnet-spinel-lherzolite nodules from Africa (Smith and Dawson, 1975, Table 1, specs. F, G, Ha).
4. Average compositions of spinel in 3 spinel

sea water. Thus, although the great bulk of the newly generated crust may eventually be recycled back into the mantle, it has provided a potential source of Si for the sedimentary cycle.

Brown (1977) and Brown and Hennessy (1978) have estimated contemporary crustal growth rates at destructive plate margins of $0.5 \text{ km}^3 \text{ y}^{-1}$. Growth is by the accretion of mantle-derived magma, although some unknown (but reputedly small) proportion of this comes from crustal recycling.

The Si content of the accreting materials varies from margin to margin and with the type and evolutionary stage of each margin (Section 1.2.1.4.3). It may be assumed that, on a global basis, it has the composition of average continental crust, i.e. $Si = 27.1$ wt. %. Calculations then suggest that 4×10^8 t of Si are added to the crust each year via destructive plate magmatism.

No recent, reliable estimate of the total volume of within-plate magmatism is available. Gass (1978) indicates that such volumes are less than those of subduction zone magmatism. This would place a maximum of 4×10^8 t y^{-1} Si transport in such settings, assuming roughly comparable average Si contents for the volcanic rocks in each broad environment.

The total figure for Si transferred from mantle to crust is approximately 9×10^9 t y^{-1}, of which perhaps 8×10^9 t y^{-1} is recycled into the mantle via subduction processes.

1.2.1.2   Behaviour of Si during mantle fusion

During the fusion process, Si is partitioned between melt and solid ($\pm$ vapour) phases. The composition of the crystalline phases must continuously change (Mysen and Kushiro, 1977), not only by changes in solid solution but by reaction between crystals and liquids to form new phases, e.g.

garnet + 0.67 diopside + 0.14 enstatite $\rightleftharpoons$ 0.22 forsterite + 1.61 liquid
(Harrison, 1979)

Little is known about the form in which Si is partitioned into the melt. Recent studies on quenched liquids (glass) having compositions relevant to magmas suggest that when chain silicates (such as pyroxenes) melt, the disproportionation may be represented by:

$3Si_2O_6^{4-} \rightarrow 2SiO_4^{4-} + 2Si_2O_5^{2-}$ (Virgo *et al.*, 1979)

Similarly, sheet structures, such as phlogopite, may break down in the melt to chains and three-dimensional network units:

$2Si_2O_5^{2-} \rightarrow Si_2O_6^{4-} + 2SiO_2^{0}$

Reactions between these units are also likely. Thus, Virgo *et al.* (1979) report that SiO_4^{4-} monomers are unstable in the presence of SiO_2^{0} network units, forming either $Si_2O_6^{4-}$ chains or $Si_2O_5^{2-}$ sheets, depending on the ratio of SiO_4^{4-} to SiO_2^0. The preliminary evidence indicates that the nature and extent of such reactions are dependent on the bulk composition of the system involved, as well as on P, T, and the activity and species of volatiles present.

1.2.1.2.1 Si distribution and mantle fusion. The compositions of the melts generated by melting of the mantle depend on several factors: the chemical composition and mineralogy of the source material, the P–T regime, the activity and compositions of any volatile phases present, whether melting is an equilibrium or disequilibrium process and whether or not the system remains closed during fusion.

Certain of these variables are now examined in the light of their effect on the Si content of the partial melts.

1.2.1.2.2 Effects of temperature (degrees of melting) on melt composition. Mysen and Kushiro (1977) have presented data on the compositional variations with degree of melting at mantle pressures of the co-existing phases of a sheared garnet lherzolite from a Lesotho kimberlite. This nodule is taken to represent the least depleted upper mantle under Southern Africa, and the experiments represent an attempt to study quantitatively the complex phase equilibria involved in the fusion of natural peridotite.

Anhydrous melting curves for the nodule at 20 kbar and 35 kbar pressure are shown in Fig. 1.2. Spinel is a likely, but unconfirmed, part of the lower temperature assemblages at 20 kbar. Though different in detail, e.g. the

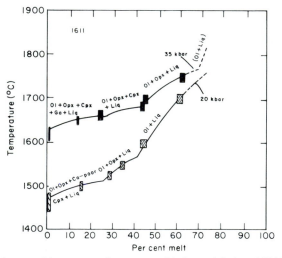

FIG. 1.2   Anhydrous melting curves of a garnet peridotite nodule (no. 1611) from a Lesotho kimberlite, at 20 kbar and 35 kbar (from Mysen and Kushiro, 1977, Fig. 3). Size of boxes includes uncertainties in temperature ($\pm10°$C) and determination of percentage of liquid ($\pm\sigma$). Ol = olivine, Opx = orthopyroxene, Cpx = clinopyroxene, Ga = garnet, Liq = liquid.

stability of garnet at high pressure, the curves show the same basic features; the notable breaks in slope as successive phases melt represent invariant reactions of the type:

$$Ol + Cpx \rightleftharpoons Opx + Liq$$ (Mysen and Kushiro, 1977, equation 3)

The concentration of Si in the liquid phase during melting of the lherzolite at 20 kbar pressure (as determined by electron microprobe analysis of quench glass) is summarized in Fig. 1.3. There is a rapid, poorly-understood, initial rise of Si in the liquid which must result, since the phase assemblage does not appear to change, from readjustments of solid solutions in the minerals. Si then remains constant at $\sim 23\%$ wt. up to 45% melting, this value being buffered by the presence of pyroxenes. Increased degrees of melting produce less Si-rich melt.

The compositions of the melt coexisting with various mineral assemblage are interpreted by Mysen and Kushiro (1977) as follows:

olivine + two pyroxenes = olivine tholeiitic
olivine + orthopyroxene = picritic
olivine alone = komatiitic

A silica-undersaturated (nepheline-normative) basaltic liquid is also recorded as melt fractions $<2\%$.

As Presnall *et al.* (1978) point out, the composition jump at the "pyroxene out" point is a misinterpretation of the phase equilibria involved. Rather, the curve should show a change of slope at this point. Nevertheless, the experiments neatly demonstrate that considerable variation in the absolute and relative abundances of Si in natural basalts may result from different degrees of partial melting of a homogeneous source.

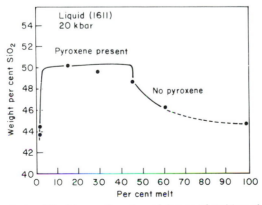

FIG. 1.3 Silica contents of liquids as a function of degree of melting of peridotite nodule 1611 (Mysen and Kushiro, 1977, Fig. 9A).

1.2.1.2.3   Effects of pressure on melt compositions. Using as starting materials synthetic compositions within the system CaO–Al$_2$O$_3$–MgO–SiO$_2$, Presnall *et al.* (1979) have determined the compositions of the first melting liquids along the solidus of simplified plagioclase and spinel lherzolites in the pressure range 1 atm. to 20 kbar. Such conditions provide a possible analogue for the generation of mid-ocean ridge basalts.

The composition of the liquids, as deduced from electron microprobe analyses of glasses in the experimental charges, are given in Table 1.4, (recalculated from Presnall *et al.*, 1979, Table 1.2), and shown in normative projections in Fig. 1.4. From 1 atm. to 9 kbar, the liquid changes from a quartz tholeiite to a tholeiite with small amounts of olivine. At high

TABLE 1.4   Electron microprobe analyses of glasses representing compositions of liquids along the solidus of simplified plagioclase- and spinel-lherzolites in the system CaO–MgO–Al$_2$O$_3$–SiO$_2$ (from Presnall *et al.*, 1979, Table 2). Estimated errors from original table omitted.

Pressure	1 atm.	7 kbar	9.3 kbar	11 kbar	14 kbar	20 kbar
Si (wt. %)	26.1	23.1	22.8	22.7	22.6	22.2
Si/Mg	3.6	2.7	2.7	2.4	2.2	1.9
Si/Ca	2.4	2.1	2.1	2.1	2.1	2.2
Si/Al	3.2	2.4	2.1	2.2	2.3	2.4

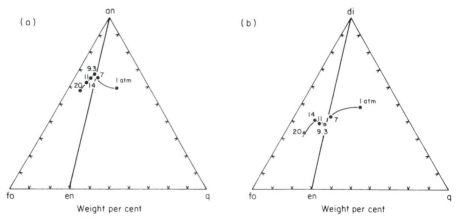

FIG. 1.4   Normative projections from diopside (A) and anorthite (B) in the system forsterite (fo)-diopside (di)-anorthite (an)-silica (SiO$_2$), showing the compositions of liquids along the solidus of a simplified lherzolite composition in the system CaO–MgO–Al$_2$O$_3$–SiO$_2$ (Presnall *et al.*, 1979, Figs 4 and 5). Numbers indicate pressures in kbar. En = enstatite, q = quartz. Note the increasingly q-poor nature of liquids as the pressure increases.

pressures, the first liquid becomes increasingly magnesian and approaches picritic affinity.

These experiments in the CMAS system do not convey the total complexity of the fusion of natural peridotites. Nevertheless, they serve to show that the compositions of the first liquids vary with pressure, not only in absolute Si abundances but, more notably, in such ratios as Si/Mg.

1.2.1.2.4 Effect of volatiles on melt composition. The importance of volatiles during the partial fusion of mantle peridotite probably varies between the various major sites of magma generation. Melting of mantle under mid-ocean ridges is thought to be essentially anhydrous (Presnall *et al.*, 1979), whilst in destructive plate margin areas, water released by dehydration reactions within the subducting lithosphere may well be critical in promoting fusion in the overlying mantle wedge. Carbon dioxide may be a factor in promoting melting beneath such diverse settings as oceanic islands and continental rift valleys.

The products of fusion of peridotite under volatile-saturated conditions may be notably different from those generated under anhydrous conditions. Thus Kushiro (1972) has shown that if fractional melting of peridotite occurs at 20 kbar under hydrous conditions, the liquid produced (A in Fig. 1.5) is of quartz-normative andesitic or dacitic composition. More recent experiments

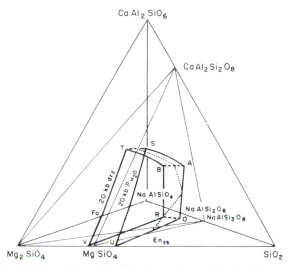

FIG. 1.5 The forsterite–enstatite$_{ss}$ liquidus boundary surface in the system Mg_2SiO_4– $NaAlSiO_4$–$CaAl_2SiO_6$–SiO_2–H_2O (Kushiro, 1972, Fig. 9). Under vapour-present conditions, the surface (U–S–A–Q) is displaced towards more silica-rich compositions relative to the anhydrous surface (V–T–B–R). A = the first liquid formed by partial melting of fosterite + enstatite$_{ss}$ + clinopyroxene + garnet + H_2O assemblages, and is of quartz-normative andesitic or dacitic composition.

(see Cox *et al.*, 1979, pp. 263–271 for review) have cast some doubt on the validity of Kushiro's data. Nevertheless a consensus remains that water-saturated melting remains a viable mechanism for the production of melts more Si-rich than basalt. Wet melting of mantle is therefore a considerably more efficient way of transferring Si from mantle to crust: enrichment factors > 1.3 may be common.

Where the mantle is fused in the presence of an excess vapour phase in which the activity of carbon dioxide is high, it is possible to generate melts with *lower* Si contents than the peridotitic sources (Mysen and Boettcher, 1975). Such magmas are markedly silica undersaturated (e.g. basanites, nephelinites, kimberlites). While they may be locally abundant, especially in continental settings and on certain oceanic islands, their volume in relation to total crust–mantle interaction is small.

1.2.1.3   Further differentiation of mantle-derived magmas

Considerable diversity in the Si contents of possible mantle melts is obviously provided by the variables described above. The situation is further complicated by the fact that magmas generated within the mantle may further differentiate during their ascent into the crust, mainly by fractional crystallization but also by wall-rock reactions and magma mixing.

1.2.1.3.1   Fractional crystallization. The behaviour of Si during the cooling crystallization of magma is controlled by the nature of the fractionating phases, which is in turn related to magma chemistry, the P–T conditions, P volatiles, etc.

Under certain conditions, non-silicate phases may be important fractionating phases from silicate magmas. Examples are the separation of spinel from basic magmas at intermediate pressures, of chromite and various sulphides in layered basic intrusions, of Fe-Ti oxides from a wide range of silicate magmas at low pressures, and of apatite from common intermediate and silicic magmas. Nevertheless, the crystallization paths of silicate magmas are dominated by the separation of silicate minerals, such that liquid trends are effectively a function of Si partitioning between solid and liquid phases.

For most magmas, such as members of the common calc-alkaline series, Si is almost invariably concentrated in lower temperature residua (Fig. 1.6, Cascades, South Aegean).

In other magma series, Si contents may show little change or actually decrease during progressive differentiation, e.g. during the early stages of the low-pressure fractionation of tholeiitic magma at low ppO_2 (Fig. 1.6, Thingmuli) and during the fractionation of trachytic magma towards phonolitic residua (Fig. 1.6, St. Helena).

Fractionation of this sort within the crust may, if highly efficient, produce Si enrichments of 1.5 to 1.6 relative to the parental magmas. In the broad

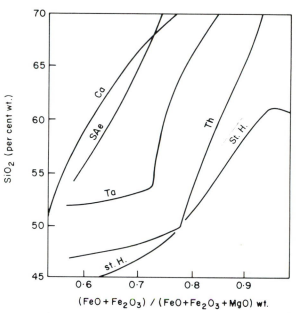

FIG. 1.6 Variation of SiO_2 contents in certain volcanic suites as a function of degree of differentiation, using $(FeO + Fe_2O_3)/(FeO + Fe_2O_3 + MgO)$ (wt.) as an index. Calc-alkaline suites—Cascades (Ca), South Aegean (SAe); tholeiitic suites—Thingmuli (Th), Talasea (Ta); alkaline suite—St. Helena (St. H.). St. Helena curve drawn from data in Baker (1969); others taken from Osborn (1976).

context, there are two main consequences of the process; firstly, it helps to widen the spectrum of igneous rock types in the crust, and secondly, there will be a tendency for the more differentiated products to reach higher crustal levels, leaving behind more mafic crystalline residua and thus contributing to crustal differentiation.

1.2.1.4 Silicon and magmatism in different tectonic settings

There is no general agreement among petrologists as to what constitutes a primary magma, i.e. one derived from its mantle source without chemical modification. Cases have been made for a range of primary magma compositions from ultramafic (MgO > 30%, akin to picrites and komatiites) to tholeiitic andesite and mugearitic compositions (MgO ~ 3–4%). Even rocks as Si-rich as trachyte have been considered to be of direct mantle origin (Bailey, 1977).

The nature of primary magmas and their subsequent differentiation histories are in part related to the tectonic environment. In the following section, simplified models of magmatism in various settings are explored in terms of Si distribution.

1.2.1.4.1 Volcanism in the ocean basins. Oceanic volcanism occurs in three settings—at the mid-oceanic ridges, along transform faults, and on oceanic islands (Gass, 1972). The latter two are volumetrically minor and, especially in the case of oceanic islands, are characterized by a wide range of differentiated products. The higher Si content of these differentiates balances to some extent the lower average Si content of the basic rocks, which are of alkaline affinity, compared to the tholeiitic products of the ridges. It may be assumed that the *average* Si concentration of associations 2 and 3 is comparable to that of the ridges and they may conveniently be omitted from the following discussion.

Current models (Gass and Smewing, 1980) envisage the generation of ocean floor basalts by the partial melting of narrow (5–10 km wide) diapirs of

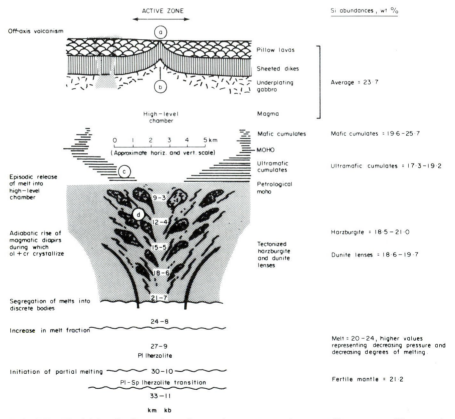

FIG. 1.7 Model for the formation of oceanic crust at active spreading centres (Gass and Smewing, 1980). Si averages and ranges for the various lithospheric units are based on data in Coleman (1977) and this paper.

mantle material rising beneath the ridges (Fig. 1.7). Fusion apparently proceeds under essentially volatile-free conditions (Presnall *et al.*, 1979). The melts migrate towards the surface but are commonly trapped in magma chambers at shallow depths below the ridge axes. There they fractionate to form a sequence of crystal cumulates, the liquid being periodically tapped to form the sheeted dyke complex of layer 3 and the pillow lava sequence of layer 2.

Such basalts are unlikely to be the direct products of mantle fusion, however. Most mid-oceanic ridge basalts have had their compositions modified to varying degrees by polybaric fractional crystallization, especially of olivine, during ascent. Estimates of the degree of fractionation vary; a range of 10–20% removal of crystalline phases is probably conservatively low. Malpas (1978, p. 542) provides a calculated example of the polybaric fractionation history of a supposedly primary ocean-floor basalt. Possible products of such fractionation events have been identified in the dunite lenses (olivine ± spinel) which occur in the harzburgitic ("depleted mantle") members of certain ophiolitic complexes (Gass and Smewing, 1980).

Making the simplifying assumption that the total spectrum of possible primary melts varies from ultramafic to mafic, the Si content of such melts will range from $\sim 20\%$ wt., equivalent to that of certain picritic basalts and komatiites, to 23.7%, the average composition of mid-oceanic ridge basalt. This latter figure was calculated by weighting the averages for Atlantic, East Pacific Rise, and Indian ocean-floor basalts given by Bryan *et al.* (1976, Table 2, cols. 1, 2, 4, and 5). This is almost identical to the average value of 23.6% Si calculated by Melson *et al.* (1976) for basalts from all oceanic spreading centres.

Using a value of $Si = 21.2\%$ wt. for fertile mantle composition (Table 1.1), these data infer that the primary magmas can show Si depletions or enrichments of $\sim 5\%$ and $\sim 15\%$ relative to the mantle. High pressure fractionation enriches the melts in Si, such that the bulk of basalts emplaced within oceanic crust show Si enrichment $> 10\%$ relative to mantle.

The mass of Si in the oceanic crust can be estimated as follows. The volumes of Layers 2 and 3 as calculated by Ronov and Yaroshevsky (1969, Table 5) are adjusted using the thicknesses of 1.30 km and 4.97 km given by Gass and Smewing (1980). Layer 2 is assumed to consist of 10% metasediment and 90% basalt, and Layer 3 to be metagabbro. The mafic rocks have $Si = 23.7$ wt. %, the sediments 19.0 wt. %, and densities of 1600 and 2900 kgm^{-3} respectively. Si in Layer 2 is then 0.263×10^{18} t and in Layer 3, 1.017×10^{18} t.

1.2.1.4.2   Petrography and mineral chemistry of ocean floor basalts. The petrography of fresh ocean-floor basalts is dominated by plagioclase, pyroxene, olivine, oxides and glass, the proportions of glass: crystals varying from 100% to O%. Bryan *et al.* (1976) have distinguished two groups of basalts on the basis of geochemistry and petrography. Group I (abyssal tholeiites) is a uniform set of rocks, depleted in large-ion lithophile elements, which is produced almost exclusively at spreading centres. Group II is more variable, though sharing the common feature of LIL-enrichment compared to Group I rocks. They have been recorded from active spreading centres, near islands and seamounts, and in fracture zones. Bryan *et al.* (1976) stress that all the basalts classed together as Group II are probably not closely related genetically. A summary of their petrography is given in Table 1.5. Other authors prefer different petrographic classifications (see discussion in Bryan, 1979) but the mineral data serve as a broad introductory guide to the mineralogy of fresh ocean-floor basalts.

Analyses of the major silicate phases and of glass in basalts from the Famous area of the Mid-Atlantic Ridge (Table 1.6, recalculated from data in Bryan, 1979) demonstrate that plagioclase + pyroxene + glass are the main repositories of Si in Layer 2 of the oceanic crust. It should be noted, however, that such mineralogy is transitory: zeolite (or greenschist) facies metamorphism will eventually be superimposed on the basalts. The open-system nature of the metamorphism will result in remobilization of silicon.

1.2.1.4.3   Magmatism at destructive plate margins. The complex relationships between the type and evolutionary stages of destructive plate margins and the character of the associated magmatism have recently been reviewed by Miyashiro (1974), Coleman (1975), and Ewart *et al.* (1977). If the term "arc system" is used in the broadest sense to include continental margin settings, it may be shown that, with increasing maturity of the arc system, the magmatism changes in character from a low-K tholeiitic type to complex associations of tholeiitic, calc-alkaline, and alkaline rocks which may show lateral zonation within the arcs.

The change in magmatic affinity is accompanied by changes in the relative proportions of the main rock types. In mature arc systems, i.e. those underlain, at least partly, by continental type crust, rocks of intermediate and silicic compositions commonly predominate over mafic varieties. A histogram (Fig. 1.8) prepared by Ewart *et al.* (1977) shows this particularly clearly. The Tonga and Kermadec segments of the arc are developed on a ridge system within an ocean basin. The Kermadecs appear from morphological evidence to represent a less advanced stage of evolution than the Tongan islands and their eruptive products are correspondingly less silicic, on average.

The Taupo arc segment, developed within continental crust, has a high proportion of rhyolitic rocks and the average composition of the eruptives is 30 wt. % Si.

In a more general review, Baker (1972) has presented estimates of the

TABLE 1.5 Petrographic characteristics of unaltered submarine basalt and basalt glass (Bryan *et al.*, 1976, Table 1).

	Group I		Group II	
	Unfractionated	Fractionated	Unfractionated	Fractionated
Plagioclase	$An_{90\,80}$: molecular Mg > Fe; K_2O < 0.02; zoning inconspicuous	$An_{80\,60}$: molecular Mg ≤ Fe; K_2O < 0.04; minor zoning	$An_{85\,60}$: molecular Mg < Fe; K_2O > 0.05; marginal zoning conspicuous	An_{60}: alkali feldspar; molecular Mg < Fe; K_2O > 0.10; prominent marginal zoning or alkali feldspar overgrowths
Olivine	$Fo_{90\,85}$; present as phenocrysts and groundmass phase and as skeletal crystals in glass	$Fo_{85\,80}$; present as phenocrysts, absent in groundmass or as quench crystals in glass	$Fo_{85\,80}$: present as phenocrysts, rare in groundmass or as quenched skeletal crystals in glass	Absent
Pyroxene	Diopsidic augite (?), detectable only in groundmass, usually as skeletal crystals	Aluminous diopsidic phenocrysts and microphenocrysts, usually optically strained and sector-zoned; may form skeletal quench crystals in glass	Titaniferous augite, rarely as phenocrysts, and prominent in groundmass	Titaniferous augite or ferro-augite as phenocrysts; distinct Ti and Al enrichment
Spinel	Aluminous, magnesian spinel; microphenocrysts associated with olivine	Absent or with distinct marginal reaction rims or magnetite overgrowths	Rare or absent	Absent
Magnetite	Absent in glass; moderately abundant as skeletal groundmass crystals	Absent in glass except as overgrowths on spinel; skeletal groundmass crystals may be abundant	Rare as phenocrysts, common to abundant in groundmass	May form euhedral phenocrysts, prominent in groundmass
Ilmenite	Absent	Absent	May be associated with magnetite in groundmass	May be associated with magnetite in groundmass; rare as phenocrysts

TABLE 1.6 Si abundances and Si:metal ratios in minerals and glasses from ocean-floor basalts: selected samples from the FAMOUS area, Mid-Atlantic Ridge. Recalculated from analyses in Bryan (1979, Tables 2, 8, 9, and 10).

Table-Spec. Nos.	Plagioclase				Pyroxene				Olivine				Glass			
	8–2B	8–3	8–5	8–13	9–1A	9–2C	9–4A	9–4B	10–1B	10–2A	10–6	10–10B	2–9	2–13	3–7	5–11
Si (wt. %)	22.3	23.1	21.8	24.4	24.7	23.0	24.1	25.0	18.8	18.8	17.6	18.4	24.0	22.8	24.5	23.8
Si/Mg	154	106	150	119	2.1	2.3	2.1	1.9	0.69	0.62	0.71	0.67	4.3	4.6	5.5	4.4
Si/Fe	65	76	100	30	5.1	3.9	5.1	4.6	1.7	2.4	1.2	1.7	3.7	2.4	3.1	3.7
Si/Ca	1.8	1.8	1.7	2.5	2.0	1.8	1.9	2.4	71	75	65	74	2.7	2.7	2.9	2.8
Si/Al	1.3	1.4	1.2	1.4	19.1	6.9	12.0	31.1	444	177	302	436	3.1	3.0	3.1	2.9
Si/Na	16.8	20.7	21.0	9.1	256	147	171	482	—	—	—	—	17.6	14.3	14.5	16.1
Si/K	894	>2305	—	734	>2468	>2295	>2412	1507	—	—	—	—	263	119	134	90
Textural type	P	P	P	P	MP	MP	P	G	P	P	MP	MP				

P = phenocryst, MP = microphenocryst, G = groundmass phase.

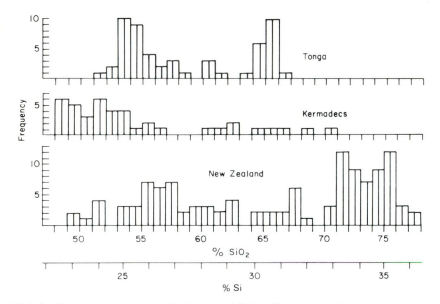

FIG. 1.8 Histogram to show the distribution of Si (wt. %) in analysed volcanic rocks from the Tongan Islands, Kermadec Islands, and Taupo volcanic zone of New Zealand (Ewart *et al.*, 1977, Fig. 5).

TABLE 1.7 Si abundance and ratios in average oceanic tholeiite (I), some calc-alkaline basalts (III), and certain island-arc tholeiites (II), (after Baker, 1972).

	I	II	III
Si (wt. %)	23.0	23.9	24.3
Si/Mg	4.8	6.4	8.1
Si/Fe	3.4	3.1	3.6
Si/Ca	2.8	3.1	3.4
Si/Al	2.6	2.5	2.5
Si/Na	11.5	16.2	10.8
Si/K	138.5	96.0	36.6

I Average oceanic tholeiitic basalt
II Average of 10 basalts from South Sandwich Islands.
III Average of 10 West Indian basalts.
Data sources in Baker (1972, Table 1)

R. Macdonald

abundances of volcanic rocks in various destructive plate margin environments and on the ocean floors (Fig. 1.9). He has also (1972, Table 1) compared the average compositions of basalts from representative settings (recalculated here, with modifications, as Table 1.7), which show that though changes in absolute Si values are slight in the sequence ocean-floor basalt to calc-alkaline basalt, the ratios of Si to various other components demonstrate marked changes.

Miyashiro (1974) has also compiled analytical data to show that with advancing development of continental-type crust, the proportion of more silicic volcanic rocks and the average Si content increase (Fig. 1.10).

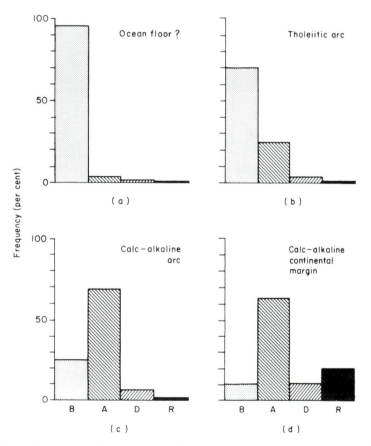

FIG. 1.9 Estimated abundances of volcanic rock types in various environments (Baker, 1972, Fig. 2). B = basalt, A = andesite, D = dacite, R = rhyolite (or nearest tholeiitic equivalents).

Petrogenetic theory of destructive margin magmatism must take into account, therefore, the type of margin involved and its stage of evolution.

1.2.1.4.4. Petrogenesis. The production of basaltic to rhyolitic volcanic rocks (and their intrusive equivalents of the gabbro-granite plutonic suite) is in some way related to plate subduction. For the andesitic products, at least, the connection might be the high H_2O contents of the subducted materials, since water-satured melting of peridotite produces melts considerably richer in Si than basalt (Section 1.2.1.2.4).

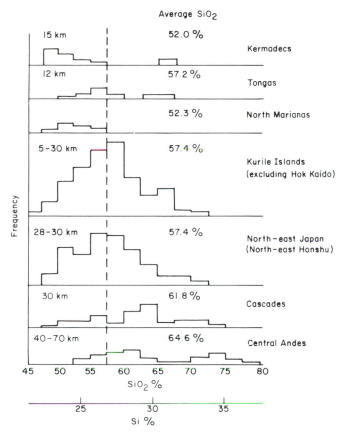

FIG. 1.10   Frequency distributions, based on available chemical analyses, of Si percentages in volcanic rocks of the tholeiitic and calc-alkaline series in island arcs and continental margins (Miyashiro, 1974, Fig. 14). Arcs and margins are approximately ordered in terms of increasing development of continental-type crust.

A multiplicity of models is available to explain their petrogenesis, including:

(i) partial fusion of subducted oceanic crust and/or mantle under varying conditions of pH_2O;

(ii) partial fusion of the overlying mantle wedge caused by the influx of water from the dehydrating, subducted crust;

(iii) generation of basalt by (i) or (ii) with subsequent fractional crystallization to produce andesitic and more silicic melts;

(iv) contamination of either mafic or andesitic melts with crustal material during magma ascent; and

(v) generation of andesites by the partial melting of lower continental crust and of more silicic magmas by melting at higher crustal levels.

Some uncertain proportion of the more silicic (Si > 30% wt.) material must be of direct mantle derivation although some has certainly been derived by crustal recycling. The necessary P–T conditions for melting of crustal materials and the composition of the liquids produced are shown in Fig. 1.11 (Brown and Hennessy, 1978). The temperatures required for melting range from 750°–900°C over a wide range of pressures. Brown and Hennessy have calculated, however, that in regions of static crust with conductive heat transfer only, Moho temperatures are only some 350°–500°C. Melting of the

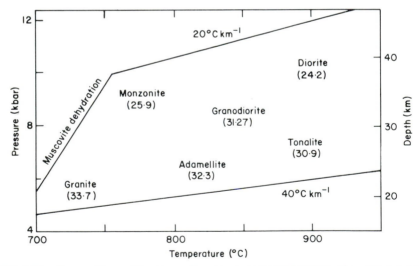

FIG. 1.11   Compositions of liquids produced by experimental melting of crustal metasediments with limiting geothermal gradients of 20 and 40°C km⁻¹ and muscovite dehydration as a lower temperature limit. Basic diagram from Brown and Hennessy (1978); Si contents (in brackets) taken from averages in Nockolds (1954).

crust seems to require the addition of heat from the mantle, probably in the form of ascending mafic and intermediate magmas. This permits, at a maximum, production of an amount of crustal melt equal to that of the heat-transporting magmas.

1.2.2 Silicon Diffusion in Crystalline and Partially Molten Mantle

A possible mechanism for the upward transport of Si through the mantle is by solid diffusion through the crystalline phases, presumably in response to P–T gradients. It appears, however, that diffusion rates of cations in silicates are extremely slow, of the order of cm in 10^9 years (Fyfe *et al.*, 1958; Foland, 1974; Fletcher and Hofmann, 1974; Hofmann and Hart, 1977; Hofmann and Magaritz, 1977). Solid diffusion cannot be considered to be a viable mechanism in mantle differentiation.

Rather more rapid rates of diffusion may be achieved where diffusion proceeds through a melt phase which wets all the grain boundaries in incipiently fused mantle rock. Hofmann and Magaritz (1977) and Magaritz and Hofmann (1978) have measured diffusion coefficients of various divalent and trivalent cations in basaltic melts and found them to range from 10^{-9} to 10^{-6} cm^2 s^{-1}. No diffusion coefficient has yet been measured for Si but it seems likely that due to its higher charge and tendency to form complexes in the melt, it will diffuse more slowly than the above values.

Characteristic diffusion distances for the measured cations will exceed 1 km only for times $> 10^9$ years. Thus while this type of diffusion may be locally important as a cause of differentiation in partially molten mantle, e.g. in the Low Velocity Zone, or at the walls of magma reservoirs, it does not contribute significantly to large-scale differentiation within the mantle (Hofmann and Magaritz, 1977).

1.2.3 Movement of Si by Migration of a Vapour Phase

The evidence for the existence of a separate gas phase in the mantle is equivocal (Yoder, 1976, pp. 78–86). If such phases do exist, the preliminary experimental data of Mysen *et al.* (1978) indicate that they would be able to migrate through the mantle, by infiltration along grain boundaries, at rapid rates (mm h^{-1}).

Experimental results by Nakamura and Kushiro (1974) suggest that up to 10% wt. Si may dissolve in the gas phase which coexists with enstatite + forsterite (simplified mantle mineralogy) at 15 kbar and temperatures $\leqslant 1315°C$ (Fig. 1.12). This represents the maximum solubility of Si, since the latter is reduced by the presence of other volatiles, notably CO_2 (Yoder, 1976).

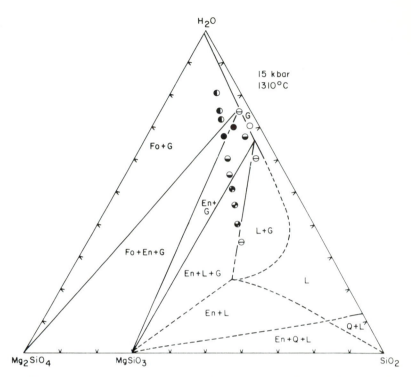

FIG. 1.12   Isothermal section of the system Mg_2SiO_4–SiO_2–H_2O at 1310°C and 15 kbar (Nakamura and Kushiro, 1974, Fig. 32). The gas phase coexisting with forsterite (Mg_2SiO_4) and enstatite ($MgSiO_3$) contains about 10% Si (wt.).

If gas streaming through the mantle is a possibility, then movement in Si in this phase seems inevitable. Probably the most suitable geological setting is at destructive plate margins. Volatiles released by dehydration reactions in the subduction zone may Si-metasomatize the overlying mantle wedge, with relative enrichment of enstatite over olivine.

Metasomatism of the upper mantle beneath uplifted, rifted regions of the stable continental shields is perhaps a ubiquitous component of alkaline magmatism in such settings. Opinions may vary, however, as to whether such metasomatism (involving such elements as Ti, Na, K, Sr, Rb, Y, Nb, Zr, Ba, the rare earth elements, F and Cl as well as H_2O) is of primary or secondary importance, a cause or a result of magma genesis (Lloyd and Bailey, 1975; Harte et al., 1975). Whichever, it appears that volatile streaming through the mantle beneath continental rifts has the ability to mobilize Si. Lloyd and Bailey (1975) report dilution and/or abstraction of Si from metasomatized mantle beneath the Rhine and East African Rift Valleys.

It is impossible to estimate the amount of Si redistributed via a fluid phase from mantle to crust in various tectonic settings. Since the final product of extensive metasomatism may be mantle fusion, its effect on Si behaviour becomes inextricably interwoven with that of magma formation.

1.3   SILICON IN THE CONTINENTAL CRUST

1.3.1   Formation of Continental Crust

Continental crust may currently be forming by several methods:

  (i) lateral accretion of andesitic island arcs;
 (ii) growth by magma addition at Andean-type destructive plate margins;
(iii) magma addition in continental interiors;
(iv) suturing of oceanic sediments and related volcanic rocks to continental masses during Himalayan type plate collisions;
 (v) underthrusting of segments of oceanic crust into deep, continental crustal levels at arc-continent margins.

These processes may be complementary; for example, the crustal thickening effected by subduction or by underthrusting may well be accompanied by magma addition from the mantle.

There is a lively, current debate about the formation of continental crust (Taylor and McLennan, 1979; Tarney and Windley, 1979, and references therein) which is not of prime concern in this paper. More germane are processes leading to the chemical differentiation of the crust.

1.3.2   Chemical Differentiation of the Crust

Whatever its ultimate origin, the continental crust is chemically stratified (Table 1.8). This may be achieved in several ways other than via the weathering cycle. Some are "primary", i.e. resulting from crust formation:

  (i) underplating of existing crust by more mafic, mantle-derived magma;
 (ii) underthrusting of continental crust by slices of oceanic crust;
(iii) emplacement at high crustal levels of mantle-derived mafic magma, resulting in local inversions of chemical stratigraphy.

Others are secondary, involving redistribution of material within existing crust:

(iv) partial melting of the lower crust and emplacement of the resultant magmas at higher levels,

TABLE 1.8   Estimates of the chemical composition (in weight per cent) of the upper, lower, and bulk continental crusts (on a volatile-free basis).

	Upper			Lower				Bulk		
	1	2	3	4	5	6	7	8	9	10
O	47.3	47.4	45.8	46.2	46.1	45.1	45.7	46.5	46.9	45.9
Si	30.7	30.9	27.7	27.8	27.2	25.1	27.1	29.0	29.8	27.0
Al	8.3	8.5	8.4	9.2	9.4	10.0	8.0	8.3	8.5	9.5
Fe	3.7	3.5	5.8	4.7	6.4	7.0	5.9	4.8	4.1	5.8
Ca	2.9	2.5	4.4	5.1	4.1	6.8	5.1	4.1	3.4	5.4
Na	2.4	2.8	2.4	2.9	1.6	2.5	2.2	2.3	3.0	2.6
K	2.8	2.7	2.2	1.6	1.3	0.5	1.7	2.4	2.2	1.3
Mg	1.4	1.4	2.4	2.1	3.1	2.5	3.2	1.9	1.7	2.1
Ti	0.4	0.4	0.5	0.5	0.7	0.5	0.7	0.5	0.4	0.5
Mn	0.1	—	0.2	0.1	0.1	—	0.2	0.1	0.1	—
P	0.1	—	0.1	—	0.1	—	0.1	0.1	—	—
Si/Mg	21.9	22.1	11.5	13.2	8.8	10.0	8.5	15.3	17.5	12.9
Si/Fe	8.3	8.8	4.8	5.9	4.3	3.6	4.6	6.0	7.3	4.7
Si/Ca	10.6	12.4	6.3	5.5	6.6	3.7	5.3	7.1	8.8	5.0
Si/Al	3.7	3.6	3.3	3.0	2.9	2.5	3.4	3.5	3.5	2.8
Si/Na	12.8	11.0	11.5	9.6	17.0	10.0	12.3	12.6	9.9	10.4
Si/K	11.0	11.4	12.6	17.4	20.9	50.2	15.9	12.1	13.5	20.8

  1. "Granitic" shell of continental crust (Ronov and Yaroshevsky, 1969, Table 5).
  2. Average "upper continental crust" (Taylor and McLennan, 1979, Table 1).
  3. "Basaltic" shell of continental crust, (Ronov and Yaroshevsky, 1969, Table 5).
  4. "Lower crust" (Smithson, 1978, Table 1).
  5. "Lower crust beneath Bournac pipe", Massif Central, France (Dupuy et al., 1979, Table 2).
  6. Average "lower continental crust" (Taylor and McLennan, 1979, Table 1).
  7. Pakiser and Robinson, 1966, Table 2.
  8. Ronov and Yaroshevsky, 1969, Table 6, col. 3.
  9. "Mean continental crust" (Smithson, 1978, Table 1).
10. "Bulk continental crust" (Taylor and McLennan, 1979, Table 1).

(v) migration of mobile elements towards higher crustal levels, with or without the presence of a free vapour phase;

(vi) large-scale tectonic movements resulting in inversions of crustal stratigraphy. Further discussion here revolves around methods (iv) and (v).

1.3.2.1   Crustal anatexis

In areas of elevated geothermal gradient, temperatures within the lower crust may be sufficiently high to cause partial fusion, especially where ascending mafic magma provides an additional heat source (Section 1.2.1.4.4). The compositional range of the melts is within the diorite → granite spectrum (Fig. 1.11), with Si characteristically being enriched in the melt relative to the residue.

Plutonism of this kind contributes towards a stratified crust, the buoyant, uprising magmas being enriched in Na, K, Th, U, Rb, and Ba as well as Si, and, arguably, leaving behind a relatively refractory lower crust of the type now represented by several Archaean granulite facies terrains (O'Hara, 1977; Tarney and Windley, 1977).

Kalsbeek (1976) provides, from Archaean rocks of the Fiskenaesset area, SW Greenland, a possible example of such anatectic fractionation of granitophile elements. There, granitic magmas, now represented by granitic gneisses, left behind a residue of hypersthene gneiss (Table 1.9). The inferred granite liquid is $\sim 3\%$ richer in Si than the parental gneisses.

Such a mechanism may be particularly potent at Andean-type plate margins, where continued uprise of basaltic to intermediate magmas from the subduction zone may trigger off a series of melting episodes within overlying crust (Brown and Hennesy, 1978; Pitcher, 1979). Each melting episode results in the relative enrichment of the granitophile elements at progressively higher crustal levels.

The process may also be particularly efficient in continental regions of high geothermal gradient, attenuated crust and high flux of mantle-derived magma, such as the Basin and Range province of the western USA, where huge volumes of silicic rocks may represent, at least in part, recycled continental crust.

TABLE 1.9 Average chemical composition of various gneiss types from the Fiskenaesset region, SW Greenland (Kalsbeek, 1976, Table 2). (e) represents the melts supposedly removed from parental rocks (c) during partial fusion, leaving gneisses (a) as residues.

	a	c	d	e
Si (wt. %)	31.4	32.2	31.9	33.6
Si/Mg	28.2	40.2	34.7	88.5
Si/Fe	10.8	14.0	12.6	34.7
Si/Ca	9.7	11.3	11.4	23.3
Si/Al	3.6	3.8	3.8	4.2
Si/Na	9.0	9.3	9.4	10.7
Si/K	32.7	25.1	21.1	10.7

(a) Average of 9 hypersthene-bearing plagioclase gneisses (Residue).
(c) Average of 13 gneisses thought to be the (Parent).
(d) Calculated mixture of 75% (a) and 25% (e).
(e) Average of 10 granitic gneisses (Extract).

Recalculated from data in Kalsbeek (1976, Table 2) on water-free basis and with all Fe expressed as FeO.

1.3.2.2   Silicon migration in volatile phases

An alternative mechanism for transporting elements from deeper to shallower crustal levels is in solution in hydrous or CO_2-rich fluids expelled from metamorphic rocks as increasingly anhydrous, high T–P mineral assemblages are developed. This mechanism has repeatedly been invoked (see Tarney and Windley, 1977) to explain the observations that certain granulite facies rocks are depleted in Th, U, K, Rb $\pm$ Cs relative to associated lower-grade rocks.

Evidence for vapour phase transport of Si in metamorphic terrains is ubiquitous. Quartz forms the most common vein filling in low- to medium-grade metamorphic rocks. Many low-grade metamorphic reactions in pelitic rocks release silica, e.g.

(i) chlorite + muscovite $\rightleftharpoons$ staurolite + biotite + quartz + vapour,
(ii) lawsonite $\rightleftharpoons$ zoisite + kyanite + quartz + vapour,

which are reactions approximating the upper limits of the greenschist and zeolite-lawsonite facies respectively (Mueller and Saxena, 1977).

The coarsening of quartz layers in low-grade pelitic rocks may also be attributable to local transport of Si through a vapour phase (Fyfe *et al.*, 1978, p. 73).

Experimental evidence on the solubility of silicates is rather scarce. Fig. 1.13 presents data on the solubility of quartz in H_2O (Holland, 1967; Fyfe *et*

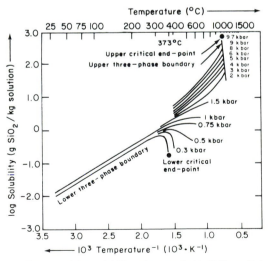

FIG. 1.13   Summary diagram of the solubility of quartz in H_2O (from Fyfe *et al.*, 1978, after Holland, 1967). Note that at the upper critical end-point, solid quartz is in equilibrium with a fluid consisting largely of SiO_2.

al., 1978). Although the form of these equilibria is in some doubt (Luth, 1976), they provide clear evidence of the ability of hydrous fluids to redistribute Si within the crust. At 600°–700°C, i.e. high-grade metamorphic conditions, about 0.5–14 g of quartz can be deposited from 100 g of solution (depending on pressure). At the lowest grades of metamorphism, this has fallen to 0.2–0.3 g quartz (Fyfe *et al.*, 1978, p. 76).

There are few published data on the solubility of the major silicates in the crust, the feldspars, over the metamorphic P–T range. Figure 1.14, constructed by Fyfe *et al.* (1978, p. 77) from the data of Currie (1968), shows that albite goes into aqueous solution incongruently, with preferential solution of Na and Si over Al. The compositional trends indicate that the degree of incongruent solution is a rather complicated function of both T and P.

The example of albite shows that continued passage of vapour through metamorphic rocks has the potential to leach Si selectively from the feldspars, producing, perhaps, as a result, such Al-phases as white micas. Morey and Hesselgesser (1951) report solubilities of albite in H$_2$O of

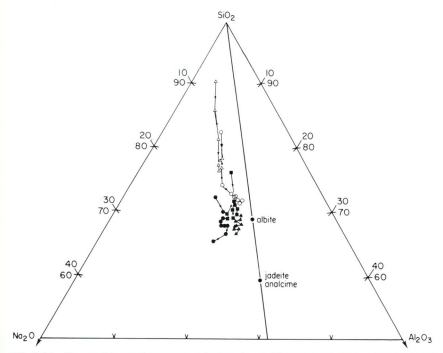

FIG. 1.14   Compositions of aqueous solutions in equilibrium with albite (in weight per cent) between 400 and 600°C and from 0.75 to 3.5 kbar. Calculated by Fyfe *et al.* (1978, Fig. 4.17) from the data of Currie (1968). ● = 400°C, ▲ = 450°C, ■ = 500°C, ○ = 550°C, △ = 600°C.

0.077 wt. % at 1 kbar and 0.267 wt. % at 2 kbar. In the natural situation, however, the amount of dissolution is dependent on the dissolution kinetics of albite. The rates may be controlled by diffusion of species from the reactant mineral through a marginal zone of the "residue" from incongruent dissolution (Fyfe *et al.*, 1978). Thus the amount of Si scavenged by the migrating fluids is controlled by the length of time over which the fluids are in contact with the reacting minerals. The critical factor then becomes the relative values of the rates of fluid flow compared with those of the mineral reactions.

Fyfe *et al.* (ibid., pp. 16–17) try to estimate the likely geological effects of this process as follows. A 30 km thick pile of sediments undergoes prograde metamorphism such that the H_2O gradient varies from 5 % at the top of the pile to 1–2 % at the base. In the T range 150°–600°C, some 3 km^3 of water per km^2 of cross-section would be evolved, which may move towards the surface something like 10^7 t of quartz per km^2 cross-section of the column.

The area of Mesozoic-Recent geosynclines has been estimated at 74.4 × 10^6 km^2 by Ronov and Yaroshevsky (1969, Table 1). If it can be assumed that turn-over time for such areas is similar to that of oceanic crust, viz. 200 Ma (Fyfe *et al.*, 1978, p. 1), then it can be calculated that 3.7 × 10^6 t y^{-1} Si are moved towards the surface by vapour phase transport. Despite the crudity of the data and of the assumptions, it would seem that such transport is not insubstantial compared to that in magmatic processes (Section 1.2.1.1).

Though, as mentioned above, geochemists have used the volatile transfer of elements to explain aspects of the chemical stratigraphy of metamorphic belts, few workers make specific mention of Si in this connection. In a detailed study of the Archaean granulite facies gneisses of the Buksefjorden region, SW Greenland, Wells (1979) suggests that the highest grade gneisses were severely depleted in Si, Na, Sr, Rb, U, and Th, which were transferred to higher crustal levels by the migration of a dispersed, H_2O-rich vapour phase. Wells (1979) mentions several modes of transport of this phase: local transport may have involved intercrystalline diffusion and infiltration through microcracks now recognizable as trails of fluid inclusions. Shear zones and pegmatite complexes may have acted as conduits for more rapid and longer-distance vapour migration.

This interpretation differs from that of Kalsbeek (1976) for the SW Greenland Archaean crust (Section 1.3.2.1), where crustal differentiation by anatexis was invoked. Wells (1979) points out, however, that both processes may act within one granulite-amphibolite terrain. Partial melting may predominate at the high temperatures of deep crustal levels ($\sim$ 60–80 km) or in areas of high pH_2O, whereas metasomatism may become important at rather higher crustal levels.

1.3.2.3 Si transport in igneous-related hydrothermal fluids

The frequent presence of quartz veins, quartz-rich pegmatites and zones of Si-metasomation near the margins of granitic plutons provides evidence of the transport of Si in the hydrous fluids expelled during the later stages of solidification of such masses.

Luth and Tuttle (1969) have presented data on the composition of the hydrous vapour phase in equilibrium with granite and granitic magmas. Though subject to considerable experimental uncertainties (Luth, 1976), the data indicate that the vapour phase in equilibrium with hydrous granite magma may carry about 8% (wt.) solid material at 10 kbar pressure. The anhydrous composition of this vapour is granitic, though rather more sodic than the coexisting magma. At temperatures 25°C below the solidus at 2 kbar pressure, the (anhydrous) vapour composition may be $\sim 96\%$ SiO_2.

Depending on the pressure and the degree of crystallization of the magma body, a range of vapour phase compositions may migrate from the magma into the country rocks. Metasomatism, formation of pegmatites, aplites, and quartz veins will variably accompany such migration. It may be noted that all these vapour-phases, and especially those at low pressures, at Si-enriched and the process must be considered to be a viable way of locally redistributing Si within the crust.

1.3.3 Mineralogy and Composition of the Continental Crust

Seismic velocity distributions within continental crust are sufficiently complex so that no single, simple model of crustal structure can be generally applicable (Wyllie, 1971; Ringwood, 1975). Smithson (1978) stresses the lateral and vertical heterogeneity of the crust but describes it in terms of three zones: an upper zone of supracrustal rocks invaded by granites, a middle zone of migmatites and a lower zone of highly variable composition, which includes supracrustal rocks, gabbroic rocks (pyroxene granulites), amphibolites, and granitic intrusives.

Tarney and Windley (1977) also note the complexity of the lower crust and list as possible materials satisfying the seismic and density constraints: (a) gabbro, (b) amphibolite, (c) mafic granulite, (d) eclogite or peridotite intercalated with sialic crust, and (e) a garnet-rich crust of intermediate composition.

Despite such complexities, most attempts to provide estimates of crustal compositions have used the almost unavoidable simplication of distinguishing an "upper" and "lower" crust. Selected estimates are given in Table 1.8. Si abundances are rather close, the range of estimates for each crustal level

R. Macdonald

being 2–3 % absolute. Ratios of Si: other elements show considerably more scatter, however, reflecting the difficulties of estimating an average composition for such a heterogeneous mass. All estimates indicate that Si and K are enriched, and the other major elements depleted or constant, in the upper relative to the lower, continental crust.

From the abundances and distribution of various rock types in large structural units of the crust, such as shields, platforms and fold mountain belts, Ronov and Yaroshevsky (1969) have calculated the average chemical compositions of the upper ("granitic") and lower ("basaltic") crust in continental and sub-continental (i.e. transitional to oceanic) settings. Their estimates of the mass and SiO_2 contents of each layer have been used to calculate the Si data given in Table 1.10. In conjunction with comparable data for the oceanic crust (Section 1.2.1.4.1), values indicate that some 80 %

TABLE 1.10 Si distribution in the upper ("granitic") and lower ("basaltic") crustal layers of the continents and transitional continental/oceanic settings (data from Ronov and Yaroshevsky, 1969).

Crustal layers	Average thickness km	Mass (10^{18} t)	Si (wt. %)	Si (Mass) (10^{18} t)
Continental				
"granitic"	20.1	8.20	29.89	2.45
"basaltic"	20.1	8.70	27.22	2.37
Sub-continental				
"granitic"	9.1	1.61	29.89	0.48
"basaltic"	11.7	2.21	27.22	0.60

TABLE 1.11 Si distribution in main petrographical types of the Earth's crust (data from Ronov and Yaroshevsky, 1969, Tables 3 and 7).

Rock type	Percentage volume of crust	Mass (10^{18} t)	Mass Si (10^{18} t)
Granites	10.4	2.95	0.99
Granodiorites, diorites	14.2	3.11	0.86
Syenites, nepheline syenites	0.4	0.11	0.03
Basalts and related rocks	42.5	12.70	2.87
Dunites, peridotites	0.2	0.06	0.01
Gneisses	21.4	5.96	1.94
Crystalline schists	5.1	1.41	0.41
Marbles	0.9	0.25	0.01
Sediments	7.9	2.01	0.52

TABLE 1.12 Si abundances and masses in major silicate minerals of crustal igneous and metamorphic rocks.

Mineral	% volume of crust	Density kg m^{-3}	Mass $\times 10^{18}$ t	Si weight per cent	Si mass $\times 10^{18}$ t
Quartz	12	2650	3.33	46	1.53
Alkali feldspar	12	2600	3.19	30	0.96
Plagioclase	39	2650	10.55	29	3.06
Mica	5	2800	1.43	17	0.24
Amphibole	5	3200	1.63	23	0.37
Pyroxene	11	3300	3.71	23	0.85
Olivine	3	3700	1.13	15	0.17
Others	13	—	3.59	—	1.10

Volume percentages from Ronov and Yaroshevsky (1969). Density and Si (wt. %) values rather arbitrarily chosen from ranges observed in crustal species of each mineral (Condie, 1974).

of Si in the crust (excluding the thin sedimentary layer) is in the continental and transitional crust.

The abundances of the main rock types in the crust as a whole are given in Table 1.11; mass and volume estimates are from Ronov and Yaroshevsky (1969, Table 7) and average compositions of rock types in Nockolds (1954) are used to calculate Si abundances.

The estimated volumes and Si contents of the most abundant minerals in the Earth's crust are given in Table 1.12, where it can be seen that some 70% of crustal Si is bound in feldspars and quartz. A more complete list of Si abundances in silicate minerals may be found in Condie (1974, Table 14-D-1).

1.3.4 Silicon Behaviour during Metamorphism

Si abundances in common metamorphic rocks are listed in Table 1.13 (from Condie, 1974). The limits are naturally defined by the metacarbonates and metaquartzites. Between these is the complete spectrum of concentrations reflecting the complex sedimentary and/or igneous processes which produced the original mineralogy of the rocks and also any subsequent metasomatic changes which they may have undergone.

In many areas it has been established that progressive regional metamorphism has been "isochemical", i.e. chemical migrations have been restricted to relatively small domains (Mehnert, 1969). Elsewhere, larger scale (m to km) movements of elements have been invoked, such transport being connected with the migration of volatile phases (Section 1.3.2.2) in the

P–T range prior to, and overlapping with, the onset of partial melting conditions.

Metamorphic reactions may proceed, therefore, under conditions of either an "open" or "closed" system as regards Si. Free SiO_2, either as quartz or dissolved in a fluid phase, may be consumed or released during these reactions. Condie (1974) provides specific examples of both types of behaviour.

TABLE 1.13 Silicon abundance in common metamorphic rocks, recalculated from data compiled by Condie (1974, Table 14-M-1).

Rock type	Mean Si (wt. %)	Range Si (wt. %)	Number of samples
Roofing slates	30.3	25.3–32.1	22
Slates	28.9		61
Slates, Precambrian	28.2	24.0–31.3	33
Phyllites	28.1		50
Phyllites, Norway	29.9		29
Quartzites	45.5	44.1–46.6	15
Metaquartzites	37.5		18
Mica schists	30.2		37
Mica schists	30.1		103
Marbles	5.8	2.4– 8.5	10
Quartzofeldspathic gneiss, Emeryville area, NY (USA)	32.2	31.8–33.1	4
Quartzofeldspathic gneiss, Colton area, NY (USA)	30.3	28.7–32.1	4
Paragneiss-paraschist, Canadian Shield	30.7	28.6–31.0	396
Banded gneiss-migmatite, Canadian Shield	30.6	27.7–31.3	2573
Quartzofeldspathic gneisses	33.1		250
Amphibolites, Emeryville area, NY (USA)	22.5	20.5–23.4	7
Pyroxene amphibolites, Colton area, NY (USA)	22.4	21.8–23.6	9
Amphibolites	23.5		200
Granulites, Musgrave Range, (Australia)	28.3		23
Granulites, Fraser Range, Australia	25.6		15
Granulites, Cape Naturliste (Australia)	31.6		74
Eclogites	21.5	19.9–23.3	7
Eclogites	22.9		34
Glaucophane schists	24.4	22.7–27.6	3

a Blank indicates ranges are not available.
b Range of mean values from areas studied in detail.

REFERENCES

Bailey, D. K. (1976). *In* "The Evolution of the Crystalline Rocks" (D. K. Bailey and R. Macdonald, eds.). Academic Press, London and New York.

Baker, I. (1969). *Bull. Geol. Soc. Amer.*, **80**, 1283–1310.

Baker, P. E. (1972). *Jl. Earth Sci.* (Leeds), **8**, 183–195.

Bickle, M. J., Hawkesworth, C. J., Martin, A., Nisbet, E. G., and O'Nions, R. K. (1976). *Nature*, **263**, 577–580.

Bishop, F. C., Smith, J. V., and Dawson, J. B. (1978). *Lithos*, **11**, 155–173.

Boyd, F. R. and Nixon, P. H. (1975). *Phys. Chem. Earth*, **9**, 431–454.

Boyd, F. R. and Nixon, P. H. (1978). *Geochim. et Cosmochim. Acta*, **42**, 1367–1382.

Brown, G. C. (1977). *Nature*, **265**, 21–24.

Brown, G. C. and Hennessy, J. (1978). *Phil. Trans. R. Soc. Lond. A.* **288**, 249–261.

Bryan, W. B. (1979). *J. Petrol.*, **20**, 293–325.

Bryan, W. B., Thompson, G., Frey, F. A., and Dickey, J. S. (1976). *J. Geophys. Res.*, **81**, 4285–4304.

Carswell, D. A. (1980). *Lithos*, **13**, 121–138.

Carswell, D. A. and Dawson, J. B. (1970). *Contr. Mineral. Petrol.*, **25**, 163–184.

Coleman, P. J. (1975). *Earth Sci. Rev.*, **11**, 47–80.

Coleman, R. G. (1977). "Ophiolites". Springer-Verlag, Berlin.

Condie, K. C. (1974). *In* "Handbook of Geochemistry", Vol. II/2 (K. H. Wedepohl, ed.). Springer-Verlag, Berlin.

Cox, K. G., Bell, J. D., and Pankhurst, R. J. (1979). "The Interpretation of Igneous Rocks". George Allen & Unwin, London.

Currie, K. L. (1968). *Am. J. Sci.*, **266**, 321–341.

Dupuy, C., Leyreloup, A., and Vernieres, J. (1979). *Phys. Chem. Earth*, **11**, 401–415.

Ewart, A., Brothers, R. N., and Mateen, A. (1977). *J. Volcanol. Geoth. Res.*, **2**, 205–250.

Fletcher, R. C. and Hofmann, A. W. (1974). *In* "Geochemical Transport and Kinetics" (A. W. Hofmann, B. J. Giletti, H. S. Yoder, Jr., and R. A. Yund, eds.). *Carn. Inst. Wash. Publ.* **634.**

Foland, K. A. (1974). *In* "Geochemical Transport and Kinetics" (A. W. Hofmann, B. J. Giletti, H. S. Yoder, Jr., and R. A. Yund, eds.). *Carn. Inst. Wash. Publ.* **634.**

Fyfe, W. S., Turner, F. J., and Verhoogen, J. (1958). *Geol. Soc. Amer. Mem.*, **73**, 259 p.

Fyfe, W. S., Price, N. J., and Thompson, A. B. (1978). "Fluids in the Earth's Crust". Elsevier, Amsterdam.

Gass, I. G. (1972). *Phil. Trans. R. Soc. Lond. A.* **271**, 131–140.

Gass, I. G. (1978). *Sci. Prog. Oxf.*, **65**, 251–268.

Gass, I. G. and Smewing, J. D. (1980). *In* "The Sea", Vol. 7 (C. Emiliani, ed.). John Wiley and Sons, New York.

Harris, P. G., Reay, A. and White, I. G. (1967). *J. Geophys. Res.*, **72**, 6359–6369.

Harrison, W. J. (1979). *Carn. Inst. Wash. Yr. Bk.*, **78**, 562–568.

Harte, B. (1978). *Phil. Trans. R. Soc. Lond. A.* **288**, 487–500.

Harte, B., Cox, K. G., and Gurney, J. J. (1975). *Phys. Chem. Earth*, **9**, 477–506.

Hazen, R. M. and Finger, L. W. (1978). *Science*, **201**, 1122–1123.

Hofmann, A. W. and Hart, S. R. (1977). *Earth Planet. Sci. Letters*, **38**, 44–62.

Hofmann, A. W. and Magaritz, M. (1977). *J. Geophys. Res.*, **82**, 5432–5440.

Holland, H. D. (1967). *In* "Geochemistry of Hydrothermal Ore Deposits" (H. L. Barnes, ed.). Holt, Rinehart and Winston, New York.

Kalsbeek, F. (1976). *In* "The Early History of the Earth" (B. F. Windley, ed.). Wiley, London.

Kushiro, I. (1972). *J. Petrol.*, **13**, 311–334.

Kushiro, I. (1973). *In* "Lesotho Kimberlites" (P. H. Nixon, ed.). Lesotho National Development Corporation, Maseru.

Liebaü, F. (1972). *In* "Handbook of Geochemistry", Vol. II/2 (K. H. Wedepohl, ed.). Springer-Verlag, Berlin.

Lloyd, F. E. and Bailey, D. K. (1975). *Phys. Chem. Earth*, **9**, 389–416.

Luth, W. C. (1976). *In* "The Evolution of the Crystalline Rocks" (D. K. Bailey and R. Macdonald, eds.). Academic Press, London and New York.

Luth, W. C. and Tuttle, O. F. (1969). *Geol. Soc. Amer. Mem.*, **115**, 513–548.

Magaritz, M. and Hofmann, A. W. (1978). *Geochim. et Cosmochim. Acta*, **42**, 847–858.

Malpas, J. (1978). *Phil. Trans. R. Soc. Lond. A.*, **288**, 527–546.

Mehnert, K. R. (1969). *In* "Handbook of Geochemistry" Vol. 1 (K. H. Wedepohl, ed.). Springer-Verlag, Berlin.

Melson, W. G., Vallier, T. L. Wright, T. L., Byerly, G., and Nelen, J. (1976). *A.G.U. Monograph Series*, **19**, 351–368.

Meyer, H. O. A. (1977). *Earth Sci. Rev.*, **13**, 251–281.

Miyashiro, A. (1974). *Am. J. Sci.*, **274**, 321–355.

Morey, G. W. and Hesselgesser, J. M. (1951). *Econ. Geol.*, **46**, 821–835.

Mueller, R. F. and Saxena, S. K. (1977). "Chemical Petrology". Springer-Verlag, Berlin.

Mysen, B. O. and Boettcher, A. L. (1975). *J. Petrol.*, **16**, 549–593.

Mysen, B. O. and Kushiro, I. (1977). *Amer. Miner.*, **62**, 843–865.

Mysen, B. O., Kushiro, I., and Fujii, T. (1978). *Carn. Inst. Wash. Yr. Bk.*, **77**, 793–797.

Nakamura, Y. and Kushiro, I. (1974). *Carn. Inst. Wash. Yr. Bk.*, **73**, 255–258.

Nixon, P. H. and Boyd, F. R. (1973). *In* "Lesotho Kimberlites" (P. H. Nixon, ed.). Lesotho National Development Corporation, Maseru.

Nockolds, S. R., (1954). *Bull. Geol. Soc. Amer.*, **65**, 1007–1032.

O'Hara, M. J. (1977). *J. Geol. Soc. Lond.*, **134**, 185–200.

O'Hara, M. J., Saunders, M. J., and Mercy, E. L. P. (1975). *Phys. Chem. Earth*, **9**, 571–604.

Osborn, E. F. (1976). *In* Intern. Cong. on Thermal Waters, Geotherm. Energy and Vulcanism of the Mediterranean Area. Vol. 3, 154–167.

Pakiser, L. C. and Robinson, R. (1966). *A.G.U. Monograph Series*, **10**, 620–626.

Pitcher, W. S. (1979). *J. Geol. Soc. Lond.*, **136**, 627–662.

Presnall, D. C., Mysen, B. O. and Kushiro, I. (1978). *Amer. Miner.*, **63**, 597.

Presnall, D. C., Dixon, J. R., O'Donnell, T. H., and Dixon, S. A. (1979). *J. Petrol.*, **20**, 3–35.

Ringwood, A. E. (1975). "Composition and Petrology of the Earth's Mantle", McGraw-Hill, New York.

Ronov, A. B. and Yaroshevsky, A. A. (1969). *A.G.U. Monograph Series*, **13**, 37–57.

Smith, J. V. and Dawson, J. B. (1975). *Phys. Chem. Earth*, **9**, 309–322.

Smithson, S. B. (1978). *Geophys. Res. Lett.*, **5**, 749–752.

Tarney, J. and Windley, B. F. (1977). *J. Geol. Soc. Lond.*, **134**, 153–172.

Tarney, J. and Windley, B. F. (1979). *J. Geol. Soc. Lond.*, **136**, 501–504.

Taylor, S. R. and McLennan, S. M. (1979). *J. Geol. Soc. Lond.*, **136**, 497–500.

Virgo, D., Mysen, B. O., and Seifert, F. (1979). *Carn. Inst. Wash. Yr. Bk.*, **78**, 502–506.
Wells, P. R. A. (1979). *J. Petrol.*, **20**, 187–226.
Wyllie, P. J. (1971). "The Dynamic Earth". Wiley, New York.
Yoder, H. S., Jr. (1976). "Generation of Basaltic Magma". National Academy of Sciences, Washington DC.

2

The Global Cycle of Silica

R. Wollast

Oceanography Laboratory
University of Brussels, Belgium

and

F. T. Mackenzie

Department of Geological Sciences
Northwestern University, USA

2.1 INTRODUCTION

A description of the global cycle of silica requires knowledge of the major agents responsible for the distribution of materials at the earth's surface. These agents must represent to a significant degree the transport paths and fluxes among the various reservoirs considered. First, we will discuss the main transport paths of dissolved and particulate silica to the ocean today in a simple three-box model involving land, atmosphere, and ocean. Particular attention will be focused on the riverine source of silica for the ocean because of its importance to the material balance. We will then examine in detail processes occurring in the ocean reservoir and present a mass balance for dissolved and particulate materials in the ocean. Finally, we will discuss the fate of silica in the sedimentary cycle over geologic time.

   The nature and origin of silicate minerals in the ocean have received a great deal of attention since the publication of Sillén (1961) on the possible control of the composition of sea water by reactions of silicate minerals with dissolved species. In 1966, Mackenzie and Garrels (1966a,b) showed that this concept was consistent with a mass balance for the oceans based on removal of river-borne cations from sea water by reactions involving detrital clays. This theory of reverse weathering has since been a matter of controversy. Thus, it is important to consider the current fluxes of the major cations Na^+, K^+, Mg^{2+}, and Ca^{2+} in an attempt to understand the silica cycle.

2.2 TRANSPORT PATHS

2.2.1 River Input

Nearly 90% of the total material entering the ocean annually is transported by streams as dissolved and suspended materials. Until recently, Livingstone's (1963) pioneering compendium of river-water analyses formed the basis for calculation of the chemical composition of average river water. However, Meybeck and Martin (1976, 1977, 1978, 1979a,b,c) have extended Livingstone's work, and that of Alekin and Brazhnikova (1960, 1962, 1968), and arrived at an improved estimate of the average composition of river waters, as well as the chemical composition of the particulate load of rivers. In this paper, we rely heavily on the work of Meybeck and Martin.

2.2.1.1 General comments

To evaluate the net influx of dissolved species to the ocean, the river load has to be corrected for sea salts transported via the atmosphere from the ocean to the continents and rained out mainly in coastal precipitation. Table 2.1 shows the average concentration of selected dissolved and particulate elements in rivers from Meybeck (1977), and the corresponding net fluxes corrected for sea-salt cycling from Martin and Meybeck (1979). The

TABLE 2.1 Elemental content and fluxes of world major rivers (adapted from Martin and Maybeck, 1979).

| | Concentration | | Net Flux | | |
	Dissolved 10^{-3} g l^{-1}	Particulate 10^{-3} g g^{-1}	Dissolved[a] 10^{12} g y^{-1}	Particulate[b] 10^{12} g y^{-1}	Total 10^{12} g y^{-1}
Na	5.1	7.1	131	110	241
Mg	3.8	13.5	129	209	338
Al	0.05	94.0	1.9	1457	1459
Si	5.42	285	203	4417	4620
K	1.35	20.0	50	310	360
Ca	14.6	21.5	495	333	828
Fe	0.04	48.0	1.5	744	746

(a) Based on a total river discharge of 37 400 km^3 y^{-1} and corrected for oceanic salts and pollution.
(b) Based on total river particulate load of 15.5×10^{15} g y^{-1}.

corrections of fluxes for cyclic salts and pollution are still debatable estimates (e.g. Holland, 1978; Maynard, 1981), and affect mainly the evaluation of the net flux of Na^+ by perhaps as much as 20%.

It can be seen in Table 2.1 that the particulate load constitutes by far the most important contribution (88%) of total river discharge of materials to the ocean. The amount carried as solids should be increased by bed load transport, which usually is considered to be about 10% of the total suspended load (Blatt et al., 1972). The mean chemical composition of river suspended matter closely approximates that of average shale (Table 2.2).

TABLE 2.2 Comparison of the average chemical composition of river suspended material, shales, and soils.

	River suspended material[a] (wt. %)	Shales[b]	Soils[c]
SiO_2	61.0	61.9	70.7
Al_2O_3	17.8	16.9	13.5
Fe_2O_3(d)	6.86	7.20	5.44
MgO	2.20	2.40	1.04
CaO	2.94	1.49	1.92
Na_2O	0.95	1.07	0.85
K_2O	2.40	3.70	1.64

(a) Calculated from the data of Meybeck (1977).
(b) After Clark (1924).
(c) After Vinogradov (1959).
(d) Total iron as Fe_2O_3.

This resemblance is expected because suspended solids in rivers are derived from shales. Sedimentary rocks constitute about 66% of the rocks exposed at the earth's surface; fine-grained rocks, like shales, comprise at least 65% of the sedimentary rock mass. Thus, roughly 50% of surface erosion products come from shaly rocks.

The mineralogy of the suspended matter carried by rivers, which is important for understanding the possible diagenetic reactions involving silicates in the ocean system, is not well documented. There are numerous analyses either of the clay fraction or of sands carried by rivers, but only a few total quantitative analyses are reported in the literature. As examples, we have evaluated the average mineralogical composition of two large river systems, the Amazon and the Mississippi (Table 2.3)

Gibbs (1967a,b) evaluated the mineralogical composition of the Amazon River at its mouth. From the mean composition of the <2 μm and 2–20 μm

TABLE 2.3   Mineralogical composition of river suspended load and shales (wt. %).

	Quartz	Plagioclase	K-feldspar	Kaolinite	Illite	Mont-morillonite	Chlorite
Amazon	15	2	2	27	31	20	5
Mississippi	30	6	4	8	13	38	—
Shales	31	←— 4.5 —→		8	15	36	2

size fractions given by Gibbs, the mean total composition was calculated assuming, given the size distribution of Gibbs, that the $<2\,\mu m$ fraction represents 75% of the suspended load. It should be noted that this total composition does not include the $>20\,\mu m$ fraction and, thus, underestimates the contribution of quartz and feldspars.

The composition of the suspended load of the Mississippi River was evaluated from the mineralogical data of Kennedy (1965) and Griffin (1962) and the chemical analysis of its suspended matter, as given by Garrels and Mackenzie (1971). It was pointed out earlier that the average chemical analysis of the suspended load of rivers corresponds very closely to the average chemical composition of shales. It is also reasonable to assume that shale is the major rock type contributing materials to the suspended load of rivers. Therefore Table 2.3 includes the mean mineralogical composition of shales for comparison with river suspended sediments. The overall average of 300 samples of shales analysed by Shaw and Weaver (1965) is 30.8% quartz, 4.5% feldspar, 60.9% clay minerals, and 7% of other non-silicate materials. According to Weaver (1967), 60% of the clay fraction in recent shales is represented by montmorillonite, 25% by illite, 12% by kaolinite, and 3% by chlorite.

Elderfield (1976) discussed in detail the distinctions that need to be taken into account when the nature and crystallography of detrital clay minerals are considered in terms of their potential to react with sea water. For instance, the weathering products of biotite and muscovite transported by rivers may be expandable products, but are generally not true montmorillonites. These minerals contain centres of negative lattice charge close to interlayers which may lead to a strong linkage of potassium to adjacent tetrahedral sites. This type of substitution is not the case for true montmorillonites. Furthermore, even if the clay fraction transported by rivers contains small proportions of clay minerals like vermiculite and glauconite, their possible reactivity with sea water may play a relatively important role. Unfortunately, a very limited amount of data is available concerning the abundance of these minerals in river suspended matter.

2.2.1.2 Weathering processes and origin of dissolved species

The principle source of silica (SiO_2) in dissolved and particulate form for the oceans is rivers and, ultimately, rocks exposed at the earth's surface. Chemical weathering of sedimentary and crystalline rocks produces dissolved and solid materials. The dissolved products of weathering enter streams directly via sheet run-off or indirectly by groundwater flow. Solid products of weathering are either the original minerals of a weathered rock or chemical alteration products, such as the clay minerals kaolinite [$Al_2Si_2O_5(OH)_4$] and montmorillonite [$Ca_{0.2}Mg_{0.34}Fe_{0.21}Al_{1.67}Si_{3.83}O_{10}(OH)_2$]. Examples of chemical reactions involving rock minerals that lead to production of dissolved and solid products of weathering are:

$$NaCl = Na^+ + Cl^- \tag{1}$$
halite

$$CaCO_3 + CO_2 + H_2O = Ca^{2+} + 2HCO_3^- \tag{2}$$
calcite

$$2KAlSi_3O_8 + 2CO_2 + 11H_2O = Al_2Si_2O_5(OH)_4 + 2K^+ + 2HCO_3^-$$
K-feldspar   kaolinite

$$+ 4H_4SiO_4 \tag{3}$$

Reaction (3) produces both dissolved silica and silica incorporated in a solid, in this case the clay mineral kaolinite. Solid materials may accumulate for some time at the site of weathering to form soils but are eventually washed into streams and transported seaward.

One means of assessing sources of dissolved constituents transported to the oceans is to consider the chemical composition of average river water. The average composition of river waters (corrected for pollutional inputs, Meybeck, 1979; and charge balance) entering the ocean is in μeq l^{-1} or μmol $l^{-1}(SiO_2)$:

SiO_2	Ca^{2+}	Mg^{2+}	Na^+	K^+	Cl^-	SO_4^{2-}	HCO_3^-	Total
173	671	281	223	33	166	175	867	2589

Meybeck calculates that about 12% of the total mass of dissolved constituents delivered to the oceans today is due to man's activities. The elements most affected are Cl, S, and Ca.

Dissolved silica (SiO_2) is about 9% of the salinity of average river water. Because the source of SiO_2 in river waters is intimately linked to the origin of other constituents, it is necessary to examine the sources of all the major dissolved components. The following balance is modelled after the works of Garrels and Mackenzie, 1971; Holland, 1978; and Meybeck, 1979; details of some of the following calculations are given in these references.

The first step in accounting for the dissolved species in river water is subtraction of the amounts of dissolved materials in precipitation. This step is particularly important in balancing NaCl, the most important component of sea aerosol washed out of the atmosphere by precipitation. After removing the precipitation component of river water, the remainder represents dissolved constituents derived from rocks, and in the case of HCO_3^-, in part from the atmosphere:

SiO_2	Ca^{2+}	Mg^{2+}	Na^+	K^+	Cl^-	SO_4^{2-}	HCO_3^-	Total
173	671	281	223	33	166	175	867	2589
	3	16	64	1	76	8		
173	668	265	159	32	90	167	867	2421

The sufficient Na^+ is now subtracted to remove all the Cl^-. This step accounts for Cl^- present in rocks as halite (NaCl), as NaCl in fluid inclusions, and as NaCl in concentrated aqueous films surrounding mineral grains in rocks. The remainder is:

SiO_2	Ca^{2+}	Mg^{2+}	Na^+	K^+	Cl^-	SO_4^{2-}	HCO_3^-	Total	NaCl Removed (µmol)
173	668	265	159	32	90	167	867	2421	
			90		90				
173	668	265	69	32	0	167	867	2241	90

Sulphate is principally derived from the weathering of $CaSO_4$ minerals (gypsum and anhydrite) in sedimentary rocks. However, some sulphate in rivers comes from the weathering of magnesium sulphate salts in sedimentary rocks and from oxidation of sulphides (primarily FeS_2, pyrite) in sedimentary and crystalline rocks. The latter process also liberates small amounts of the cations Ca^{2+}, Mg^{2+}, Na^+, K^+ by reactions like:

Step 1. Oxidation of pyrite
$$2FeS_2 + 15/2O_2 + 4H_2O = Fe_2O_3 + 4SO_4^{2-} + 8H^+; \qquad (4)$$

Step 2. Weathering of rock minerals by "H_2SO_4"
$$\underset{\text{calcite}}{2CaCO_3} + H_2SO_4 = 2Ca^{2+} + 2HCO_3^- + SO_4^{2-}, \qquad (5)$$

and

$$\underset{\text{K-feldspar}}{2KAlSi_3O_8} + H_2SO_4 + 4H_2O = \underset{\substack{\text{montmorillonite-like}\\\text{mineral}}}{Al_2Si_4O_{10}(OH)_2}$$
$$+ 2K^+ + SO_4^{2-} + 2H_4SiO_4. \qquad (6)$$

Notice that in step 2, the weathering of an aluminosilicate mineral by the aggressive attack of "H_2SO_4" waters may result in the release of dissolved silica, as well as cations. The ratio of silica to cations released depends on the composition of the primary mineral weathered and its alteration product. The weathering of a carbonate mineral gives cations and bicarbonate.

About 95% of the sulfur as SO_4^{2-} in river waters from rock weathering is derived from sedimentary rocks; the remainder from crystalline rocks (Meybeck, 1979). Of the total sulphur in river waters derived from rock weathering, about 30% comes from the weathering of sulphides in sedimentary and crystalline rocks, 25% from sediments, and 5% from crystallines (Holland, 1978). Using these proportions, the Na/K and Ca/Mg ratios of cations accompanying sulphate oxidation in rocks (Meybeck, 1979), and the assumptions that all the Na^+ and K^+, 15% of the Ca^{2+}, and 50% of the Mg^{2+} are derived from silicate mineral weathering reactions giving an *average* molar ratio of SiO_2 to HCO_3^- of 1:1 (Garrels and Mackenzie, 1971; Holland, 1978), the following balance is obtained:

SiO_2	Ca^{2+}	Mg^{2+}	Na^+	K^+	Cl^-	SO_4^{2-}	HCO_3^-	Total	Reactants removed
173	668	265	69	32	0	167	867	2241	
14	41	22	5	3		46	25		
159	627	243	64	29	0	121	842	2085	14 silicate, 25 carbonate, 12 FeS_2

This is a tentative balance because of a number of poorly known estimates, but it does illustrate the link between sulphide mineral weathering and resultant acid attack on carbonate and silicate minerals.

The remaining sulphate comes from calcium and magnesium sulphates in sedimentary rocks in the proportion of $3Ca^{2+}$ to $1Mg^{2+}$; thus, the remainder is:

SiO_2	Ca^{2+}	Mg^{2+}	Na^+	K^+	Cl^-	SO_4^{2-}	HCO_3^-	Total	$(Ca, Mg)SO_4$ removed
159	627	243	64	29	0	121	842	2085	
	91	30				121			
159	536	213	64	29	0	0	842	1843	60

The constituents left come from the weathering of silicate and carbonate rocks by CO_2-charged soil and ground waters. From the data of Garrels and Mackenzie (1971), Holland (1978), and Meybeck (1979), it is estimated that the weathering of carbonate rocks supplies 85% of the remaining Ca^{2+} and 50% of the remaining Mg^{2+} according to a reaction of the type:

$$CaCO_3 + CO_2 + H_2O = Ca^{2+} + 2HCO_3^- \tag{7}$$

A similar reaction holds true for the magnesium component of carbonate rocks. The balance now is:

SiO_2	Ca^{2+}	Mg^{2+}	Na^+	K^+	Cl^-	SO_4^{2-}	HCO_3^-	Total	Carbonate removed
159	536	213	64	29	0	0	842	1843	
	456	107					563		
159	80	106	64	29	0	0	279	717	282

The molar ratio of Ca^{2+} to Mg^{2+} in the world's rivers calculated as carbonate removed is about 4:1. This ratio is slightly greater than that obtained for North America of 3:1 (Garrels and Mackenzie, 1971) and that given by Meybeck (1979) of 3:1 for the world. However, it is the same ratio calculated by Holland (1978) for the world's rivers and found in the average limestone composition of Clarke (1924). It is slightly less than the average ratio in continental carbonate rocks, 5:1 (Ronov, 1980), the most important source of dissolved constituents in rivers. A Ca:Mg ratio of 4:1 indicates that the ratio of limestone to dolomite weathered is about 3:1 for the world, which is certainly a reasonable estimate. The total amount of HCO_3^- now removed in carbonate minerals is 588 μeql^{-1}; about one half of this amount comes from the atmosphere, the remainder from carbonate rocks.

Now we must attempt to determine the source of the rest of the SiO_2 in river waters. This SiO_2 comes from the weathering of silicate minerals by CO_2-charged waters, reactions which also account for the remaining cations and HCO_3^-. The HCO_3^- is derived from two major sources: atmospheric CO_2 that has interacted directly with silicate minerals or has been incorporated into plants by photosynthesis, later to be released by plant decay or root respiration; and CO_2 produced by oxidation of fossil carbon already present in rocks (Garrels and Mackenzie, 1971, 1972; Holland, 1978).

The budget can be balanced by consideration of a variety of silicate reactions producing different ratios of cations: HCO_3^- :SiO_2. The simplest approach was that of Garrels and Mackenzie (1971) in which they considered the reactions of feldspars with CO_2-charged water to produce kaolinite and a montmorillonite-type mineral:

$$2\underset{\text{albite}}{NaAlSi_3O_8} + 2CO_2 + 11H_2O = \underset{\text{kaolinite}}{Al_2Si_2O_5(OH)_4} + 2Na^+ + 2HCO_3^-$$
$$+ 4H_4SiO_4, \quad (8)$$

and

$$2NaAlSi_3O_8 + 2CO_2 + 6H_2O = \underset{\text{"montmorillonite"}}{Al_2Si_4O_{10}(OH)_2} + 2Na^+ + 2HCO_3^-$$
$$+ 2H_4SiO_4 \quad (9)$$

(similar reactions can be written for K-feldspar, $KAlSi_3O_8$).

Notice that in these reactions the ratio of SiO_2 to HCO_3^- varies from $1:1$ to $2:1$. The weathering of montmorillonite to kaolinite also produces one molecule of SiO_2 for each molecule of HCO_3^-. However, weathering of the anorthite $(CaAl_2Si_2O_8)$ component of feldspars to kaolinite produces no dissolved silica:

$$CaAl_2Si_2O_8 + 2CO_2 + 3H_2O = Al_2Si_2O_5(OH)_4 + Ca^{2+} + 2HCO_3^-,$$
anorthite

$$(10)$$

and weathering of chlorite to kaolinite produces little SiO_2 relative to HCO_3^-,

$$Mg_5Al_2Si_3O_{10}(OH)_8 + 10CO_2 + 5H_2O = Al_2Si_2O_5(OH)_4 + 5Mg^{2+}$$
chlorite
$$+ 10HCO_3^- + H_4SiO_4. \qquad (11)$$

Because it is likely that most of the Na^+ and K^+ in river waters derived from silicates comes from weathering reactions like those involving feldspars, and most of the Ca^{2+} and Mg^{2+} stems from reactions like those above for anorthite and chlorite, we will use the stoichiometry of these reactions to complete the balance. First, we remove the remaining Ca^{2+} and Mg^{2+} according to reactions (10) and (11):

SiO_2	Ca^{2+}	Mg^{2+}	Na^+	K^+	Cl^-	SO_4^{2-}	HCO_3^-	Total	Silicate removed
159	80	106	64	29	0	0	279	717	
11	80	106					186		
148	0	0	64	29	0	0	93	334	11

A balance for the remaining materials can be obtained by utilizing about 75% of the remaining silica in the chemical weathering of feldspars to kaolinite, and the rest in reaction of feldspars to montmorillonite:

Feldspars to kaolinite

SiO_2	Ca^{2+}	Mg^{2+}	Na^+	K^+	Cl^-	SO_4^{2-}	HCO_3^-	Total	Feldspars removed
148	0	0	64	29	0	0	93	334	
111			←— 56 —→				56		
37	0	0	←—37—→		0	0	37	111	56

Feldspars to montmorillonite

SiO_2	Ca^{2+}	Mg^{2+}	Na^+	K^+	Cl^-	SO_4^{2-}	HCO_3^-	Total	Feldspars removed
37	0	0	←—37—→		0	0	37	111	
37			←—37—→				37		
0	0	0	←— 0 —→		0	0	0	0	37

The ratio of silica used for the two weathering reactions of 3:1 is that required by the stoichiometry of the reactions; interestingly, it is also the ratio of dissolved silica loads derived from tropical and temperate regions (3.2:1; Meybeck, 1979). In general, tropical weathering of silicate minerals produces kaolinite as the principle alteration product, whereas temperate weathering results in formation of montmorillonite (e.g. Loughnan, 1969).

Although a balance has been obtained which seems consistent with present ideas, it is not an unique representation. However, the balance gives a rough approximation of the relative contributions of different rock types to chemical weathering and sources of dissolved constituents in river water. If we compute the weight percentages of rock types dissolved from the moles of minerals weathered, the dissolved load of the world's rivers transported to the oceans is derived from the following sources: 7% from beds of halite and salt disseminated in rocks, 10% from gypsum and anhydrite deposits and sulphate salts disseminated in rocks, 38% from limestones and dolomites, and 45% from weathering of one silicate to another.

The composition of the average plagioclase feldspar weathered is $Na_{0.65}Ca_{0.35}Al_{1.35}Si_{2.65}O_8$, andesine, and is similar to the composition of plagioclase feldspar found in granodiorite, a reasonable average igneous rock composition supplying much of the Na^+ and Ca^{2+} in rivers derived from silicate weathering. The molar ratio of $Na^+:Ca^{2+}$ in river waters derived from weathering of plagioclase feldspar is essentially that found in the average plagioclase feldspar. Of the HCO_3^- in river water, 56% stems from the atmosphere, 35% from carbonate minerals, and 9% from the oxidative weathering of fossil organic matter. Reactions involving silicate minerals account for about 30% of river HCO_3^-.

This balance accords to silicate mineral weathering a greater importance than that given by Meybeck (1979), reflecting the authors' opinion that 10–20% of the Ca^{2+} and 40–60% of the Mg^{2+} in river waters is derived from silicate weathering. Meybeck assigns a trivial portion of Ca^{2+} and Mg^{2+} in rivers to silicate weathering; thus, he arrives at a ratio of silicate to carbonate weathered much lower than our estimate. Our value is more in accord with Holland's (1978) balance for the world's rivers. Inclusion of the pollution component of river waters in the balance would modify slightly the calculated ratios of rock types weathered.

In conclusion, it should be noted that the dissolved silica and cations brought to the ocean by streams must be removed, if the ocean is a steady-state reservoir. Also, the anions, Cl^- and SO_4^{2-}, must enter into chemical reactions for the same reason. Bicarbonate should be removed by reactions resulting in return of CO_2 to the atmosphere. During the past two decades controversy has raged over the steady-state ocean concept, the mechanisms and fluxes responsible for element removal from the ocean, and the time scale for removal of these elements and return of CO_2 to the atmosphere, i.e. early or late diagenesis or metamorphism.

2.2.2 Atmospheric Transport

Wind stress on the land surface results in advection of soil into the atmosphere; the dust later falls on the sea surface as particulate matter in rain or as dry fallout. Buat-Menard and Chesselet (1979) evaluated fluxes to the sea surface of various elements in particulates for the North Atlantic open ocean and found for Al a flux of 5×10^{-6} g cm^2 y^{-1}. For a mean Al content in soils of about 7%, a total flux of solid is obtained of 7×10^{-5} g cm^2 y^{-1}. If this flux is representative of all the ocean surface area (360×10^{16} cm^2), the amount of solid transferred from the continents to the ocean via the atmosphere is estimated as 2.5×10^{14} g y^{-1}. This value is within the range of 0.5–5×10^{14} g y^{-1} that we reported earlier (Mackenzie and Wollast, 1977. See also Chapter 3, Section 3.6).

The input from the atmosphere is small compared to the suspended load of rivers of 155×10^{14} g y^{-1}. However, as was pointed out by Buat-Menard and Chesselet (1979), the relative importance of this source of particulates increases for the open ocean where the contribution of river materials is reduced substantially. For example, the occurrence of quartz in deep-sea sediments of central parts of the oceans (e.g. Windom, 1976) has been related to quartz-bearing dust transported from arid regions of the continents by prevailing winds. Similarities in the distributions of feldspar and quartz in these sediments, and the mineralogy of dust samples collected over the open ocean support the conclusion that alkali feldspars also have an eolian origin in certain regions of the ocean.

The major element content and mineralogy of air-borne particles reflect closely those of continental soils and shales, although atmospheric particulates also include materials of oceanic origin (Delaney *et al.*, 1967) and show considerable enrichments in some trace metals (Buat-Menard and Chesselet, 1979). The average composition and mineralogy of shales (Tables 2.2 and 2.3) were chosen to represent the properties of dust transported from the continents to the ocean.

2.2.3 Submarine Basalt—Seawater Reactions

2.2.3.1 Hydrothermal reactions at mid-ocean ridges

There now exists excellent evidence that extensive hydrothermal circulation occurs within mid-ocean ridges and that sea water is the metasomatic fluid (e.g. Wolery and Sleep, 1976). Numerous authors have described serpentinites and greenstones collected from the sea-floor, resulting from the reaction of sea water with mid-ocean ridge basalts. Typical mineral assemblages found in hydrothermally-altered basalts consist of zeolites, montmorillonite and mixed-layer chlorite-montmorillonite. Also, several laboratory experiments have been conducted to evaluate the changes of composition of sea water during hydrothermal reactions with basalt, glass, and crystalline rocks (e.g. Holland, 1978).

The principle exchange at *high temperature* involves a release of silica, K^+, and Ca^{2+} to sea water and an uptake of Na^+ and Mg^{2+}. However, evaluation of the fluxes resulting from these hydrothermal reactions is still hazardous because of uncertainties concerning the *in situ* temperature of reaction and the volume of sea water currently cycling through mid-ocean ridges. These problems are reflected in the large discrepancies between various estimates of fluxes published in the literature (Hart, 1973; Maynard, 1976; Wolery and Sleep, 1976; Holland, 1978; Wolery, 1978; Edmond *et al.*, 1979; Wolery and Sleep, 1981). Three recent estimates (Table 2.4) are in relative good agreement within an order of magnitude. The direction and magnitude of the Na flux is obviously a problem.

An initial estimation was made by Holland (1978) who calculated fluxes assuming a cycle rate of 1.0×10^{17} g sea water y^{-1} and a maximum temperature of 250°C. The concentrations of the various elements used to calculate fluxes were derived from laboratory observations and the composition of Icelandic brines. Edmond *et al.*'s (1979) estimation was based on chemical analyses of waters from areas of submarine vents located on relatively young crust in the Galapagos region. The ^{3}He anomaly was used to obtain a value for total input to the ocean of hydrothermal fluids from vents.

Comparison of the estimates (Table 2.4) of Holland and Edmond *et al.* for hydrothermal fluxes with river fluxes of dissolved substances indicates that the removal of Mg and Na by hydrothermal processes is comparable to the

TABLE 2.4   Fluxes owing to hydrothermal reactions (in units of 10^{12} g y^{-1}).

	Holland (1978)	Edmond *et al.* (1979)	Wolery and Sleep (1976; 1981)
Na	− 100	variable	+ 11
Mg	− 130	− 180	− 61
Si	+ 28	+ 88	+ 25
K	+ 90	+ 49	− 5
Ca	+ 120	+ 130	+ 72
Fe*	−	−	+ 15

+ input to oceans; − efflux from oceans.

* Authors' estimate based on hydrothermal flux of Si and average Fe/Si ratio of sediments resulting from hydrothermal processes.

TABLE 2.5 Fluxes owing to submarine weathering of basalt (in units of 10^{12} g y^{-1}).

	Maynard (1976)	Wolery and Sleep (in press, 1981)	Mean value (this work)
Na	0*	−1 to −8	−5
Mg	0*	+7 to +46	+27
Al	0	—	0
Si	+18	+3 to +13	+10
K	−8	−4 to −20	−10
Ca	0*	+10 to +50	+30
Fe	+7	—	+7

+ input to ocean; − efflux from ocean.
* Maynard assumes that all Mg goes into new chlorite, Ca into calcite and Na into smectite.

global river input of these elements. Also, the inputs of Si, K, and Ca at mid-ocean ridges are comparable to the river dissolved load. These fluxes in our opinion are overestimated. Both estimates are based on observations of hydrothermal reactions at rather high temperature and on the assumption that the total volume of sea water circulated through mid-ocean ridges reacts under the same environmental conditions. Wolery (1978) showed that if low temperature, alteration reactions also are taken into account, net fluxes resulting from basalt–seawater reactions may be modified substantially and reduced for most components. Whatever the case, it is apparent that better estimates of hydrothermal fluxes are needed to improve the global balances of the major elements.

2.2.3.2 *Low temperature basalt—seawater reactions*

Low temperature alteration of basalt in the oceanic crust has been evaluated on a global basis by Hart (1973), Maynard (1976) and Wolery and Sleep (1976; 1981). The values reported by Hart are at least an order of magnitude higher than other estimates because Hart used in his calculations what now appears to be too large a thickness for weathered basalt crust. Thus, we have selected the more recent and independent values of Maynard and Wolery and Sleep (Table 2.5) which are in reasonably good agreement. However, the uncertainties in our adopted values are still large, and these values should be considered only as tentative estimates.

2.2.4 Fluxes Related to Sediment Pore Waters

Diagenetic processes occurring in marine sediments may affect significantly the composition of interstitial waters and induce concentration gradients responsible for exchange of dissolved species with the overlying water column. Any evaluation of fluxes associated with these processes is limited to a large extent by artifacts produced by temperature and pressure effects on pore-water composition during the sampling procedure, by lack of knowledge of concentration gradients near the sediment–water interface, and by uncertainties in the values of the effective diffusion coefficients. However, it is important to consider these potential fluxes in some detail because they may represent a significant source or sink of dissolved materials for the ocean. Also, an evaluation of these fluxes can provide some insight into diagenetic processes occuring in the marine environment. We will first consider in detail the potential silica release from sediments and then briefly evaluate fluxes of major cations associated with silicate–seawater reactions.

The increase of dissolved silica in pore waters of marine sediments is a well established general trend and the concentration gradient across the sediment–water interface must inevitably result in addition of dissolved silica to the oceans. The concentration profiles of dissolved silica are relatively well documented and because of the large concentration increase of silica in sediment pore waters with respect to overlying waters, silica concentration gradients at the interface may be estimated with good accuracy. Various regional or global estimates of silica fluxes are summarized in Table 2.6. One source of uncertainty lies in the choice of the diffusion coefficient, D, used in the calculations, which varies from 1.5 to 4×10^{-6} cm^2 s^{-1} depending on what factors, such as porosity, tortuosity, physical or biological advection, were taken into account in evaluation of the constant. Direct laboratory

TABLE 2.6   Flux of dissolved silica from pore water to the oceanic water column (in units of 10^{-5} g Si cm^{-2} y^{-1}).

Reference	Region	Flux*	D* $(10^{-6}$ cm^2 s$^{-1})$	Flux (authors' estimate)
Bischoff and Sayles (1972)	East Pacific Rise	1.8	1.5	3.7
Hurd (1973)	Central Equatorial Pacific	11.5	2.9	11.9
Fanning and Pilson (1974)	North Atlantic	7.0	3.3	6.4
Schink *et al.* (1975)	North Atlantic	9.0	4.0	6.8
This study	North Atlantic	—	—	7.2
This study	South Atlantic	—	—	2.3
This study	Atlantic-Indian Ridge	—	—	10
Heath (1974)	Global	5.5–11	1.5–3	11
Wollast (1974)	Global	4.4	2.5	5.3
This study	Global	—	—	5.1

* Authors' estimations of flux and value of the diffusion coefficient.

measurements of the flux of dissolved silica from Atlantic sediment cores (Fanning and Pilson, 1974) give a value of 3.3×10^{-6} cm^2 s^{-1} at 5°C. To evaluate the global flux of dissolved silica, we corrected the existing data by selecting a constant value of D equal to 3×10^{-6} cm^2 s^{-1} and included fluxes calculated from the pore water data recently obtained by Sayles (1981), assuming a linear gradient between 0 and 50 cm depth, to arrive at an average value. The weighted average flux for all the oceans computed from the regional fluxes is 5.1×10^{-5} g Si cm^{-2} y^{-1}, which gives a global flux of 185×10^{12} g Si y^{-1}.

Evaluation of the fluxes of major cations associated with sediment–pore–water reactions is more difficult because of small changes in concentrations of these constituents relative to sea water and to uncertainties related to temperature-induced artifacts. Our estimation of these fluxes is based primarily on the abundant data collected during the Deep Sea Drilling Project and summarized by Manheim and Sayles (1974), Sayles and Manheim (1975), and Manheim (1976). The fluxes reported by these authors (Table 2.7, column 2) are based on the assumption of a linear gradient between 0 and 1 m and are most likely underestimated because the general trends of the concentration profiles of cations in pore waters fit more closely an exponential function. The uptake of sodium given by Manheim as "-14×10^{13} g y^{-1}" is particularly suspect because of the inadequacy of the data and the uncertainty of the correction to be applied for temperature of extraction. Thus, we recalculated the flux of Na$^+$ from the data, assuming than an ionic charge balance must be maintained; this is the value given in Table 2.7.

Recent utilization of *in situ* sampling techniques and of differential chromatography as an analytical method (Sayles, 1979) provides more reliable data and more detailed concentration profiles in the upper 100 cm of sediment. However, the existing data are restricted mainly to the Atlantic Ocean. The pattern of diagnetic reactions observed by Sayles is virtually

TABLE 2.7 Flux of major cations across the sediment–water interface (in units of 10^{12} g y^{-1}).

	Manheim (1976)	Sayles (1979)	This study	Burial	Net flux
Na$^+$	-14*	$+101$	$+25$	-36	-9
Mg^{2+}	-60	-126	-88	-4.5	-93
K$^+$	-28	-78	-25	-1.3	-26
Ca^{2+}	$+30$	$+250$	$+292$	-1.4	$+291$

$-$ uptake by sediments from water column; $+$ input to water column.
* recalculated by present authors.

invariant throughout the region of the Atlantic studied: uptake of Mg^{2+} and K^+ occurs at every station investigated and release of Ca^{2+}, Na^+, and silica is equally consistent. An estimate of world average fluxes from Sayles' work is given in Table 2.7, column 3. For weighting purposes, Sayles assumed that 30 % of the sea-floor, termed margin areas, is characterized by large fluxes and is strongly influenced by continentally derived materials.

From the data reported by Sayles and Manheim, we estimated fluxes typical for various sediment types: terrigenous materials, pelagic clays, carbonates and siliceous oozes. We then integrated these fluxes over the areas of sea floor occupied by these sediment types. The results of these calculations are given in Table 2.7, column 4. Maynard (1981) has suggested that a correction should be applied to these values for sea water lost from the ocean by permanent burial of pore water. For an annual sediment mass of 20×10^{15} g deposited with an average final porosity of 30 %, 3.4×10^{12} litres of sea water are removed. The resultant fluxes for the major cations owing to burial are given in Table 2.7, along with our final calculated net flux values for cations. It appears that the correction for seawater burial is important only in the case of Na^+.

2.3 MASS BALANCE FOR DISSOLVED AND PARTICULATE SUBSTANCES IN THE OCEANS

2.3.1 Introduction

There are several different ways to approach the cycles of elements in the oceanic system depending on the choice of boundary conditions. For example, discussion can be restricted to the water column bounded by continents, atmosphere, and the sediment–water interface. In this case, the various fluxes between the sea-floor and the water column must be defined. Another approach is to include the marine sediments in the oceanic system and to evaluate from the sedimentary record the element fluxes and uplift rates of marine sediments. Furthermore, the water column could be subdivided into a photic zone, intermediate, and deep waters; and the sedimentary column into an early diagenetic layer, as distinguished from deeply buried sediments where slow rates of diagenetic reactions obtain. In fact, the literature is confused on the subject of element cycle models because of the lack of definition of boundary conditions and fluxes used in these models.

Various fluxes and processes considered in the following discussion are represented schematically in Fig. 2.1. Steady-state conditions require that the mass balance for each element is fulfilled for the entire system and also separately for the water and sedimentary columns. To write these mass

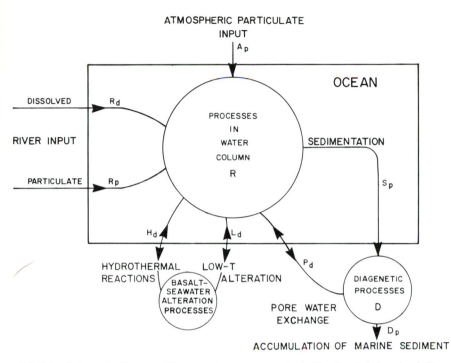

FIG. 2.1   Schematic diagram of fluxes and processes evaluated for the global cycle of silica. R_d, R_p, S_p, D_p, H_d, L_d, and P_d are silica fluxes related to riverine dissolved and particulate matter transport, oceanic sedimentation and accumulation, basalt–seawater hydrothermal and low temperature alteration reactions, and pore-water exchange, respectively; d refers to dissolved flux, p to particulate. R and D are annual amounts of silicon transferred between the solid and the aqueous phase.

balances, we must consider the global rate of the reactions occurring in the two subsystems. Therefore, we define R and D as the annual amount of a given element transferred from the solid phase to the aqueous phase or vice versa. The fluxes are taken as positive if there is a net input to sea water and negative if there is a net output.

2.3.2   Mass Balance for the Water Column

2.3.2.1   *Dissolved substances*

We will consider first the mass balance for dissolved species in the water column. The river input already has been corrected for atmospheric salt cycling and pollution, and thus, to maintain the concentration of an element constant in sea water, the net flux resulting from $R_d + H_d + L_d + P_d$ (Fig. 2.1) must be balanced by precipitation or dissolution reactions in the water column:

$$R_d + H_d + L_d + P_d = R$$

with $R > 0$ for precipitation and $R < 0$ for dissolution.

Table 2.8 summarizes the various fluxes discussed previously and in column 4 gives the computed value of R. The net fluxes for all the elements are positive and if a steady state is assumed, they must correspond to removal reaction rates owing to physico-chemical or biological processes.

TABLE 2.8   Mass balance for dissolved species in the oceanic water column (in units of $10^{12} \, g \, y^{-1}$).

	River input	Hydrothermal reactions	Submarine weathering	Pore water	Net balance (R)
Na	$+131$	$+11$	-5	$-9(?)$	$+128(?)$
Mg	$+129$	-61	$+27$	-93	$+2$
Al	$+1.9$	—	—	—	$+1.9$
Si	$+203$	$+25$	$+10$	$+185$	$+423$
K	$+50$	-5	-10	-26	$+9$
Ca	$+495$	$+72$	$+30$	$+29$	$+626$
Fe	$+15$	$+15$	$+7$	—	$+23.5$

$+$ input to ocean; $-$ efflux from ocean.

2.3.2.1.1   Cation exchange processes. A well-documented removal process is that of cation exchange with clay minerals. Suspended matter carried by rivers is rapidly involved in this process during the mixing of fresh water with sea water in the estuarine zone or in the nearby coastal zone. The process involves predominantly the clay fraction of the suspended material. Typically, the exchange positions in montmorillonites and vermiculites present in suspension in river water are dominated by Ca^{2+}, whereas degraded micas are characterized by the presence of H^+ in exchange positions. As indicated by equilibrium considerations, these cations are replaced by major cations present in abundance in sea water, mainly Na^+, Mg^{2+}, and K^+. To evaluate quantitatively this cation-exchange process on a global scale, we have used the study of Russell (1970) on the Rio Ameca River (Mexico) and those of Sayles and Mangelsdorf (1977, 1979) on the Amazon River. Both studies enable us to evaluate the extent of exchange of Na^+, K^+, Ca^{2+}, and Mg^{2+} per gram of suspended matter and the total exchange capacity of this material (Table 2.9). The general trends observed for the two river systems are similar, and the differences for each cation reflect, as would be expected, variations in the mineralogical composition of the clay fraction. The clay fraction of the Rio Ameca contains approximately 60% montmorillonite, 40% kaolinite, and 5% illite (as given by Russell, 1970), whereas Amazon clays are composed of 24% montmorillonite, 33% kaolinite, and 37% illite. Both analyses are significantly different from the average composition of shales used by us to represent the composition of suspended matter carried by rivers to the ocean (Section 2.2.1.1).

To obtain values for fluxes associated with cation exchange reactions representative of the global importance of this process, we used data concerning the equilibration of various clay minerals with sea water during laboratory experiments (Carroll and Starkey, 1960; Sayles and Mangelsdorf, 1977) and applied these values of exchange capacity and exchangeable fraction for each cation to the mean normative mineralogical composition of riverine suspended matter (see Table 2.15). Details of the normative calculation are discussed later. The results of these calculations are presented in Table 2.9.

It should be pointed out that early reactions other than reversible cation exchange may result in ions bound to clay lattices so firmly that they are unable to exchange. This irreversibility is reflected in a decline of cation exchange capacity of clay after a short period of exposure to sea water (Carroll and Starkey, 1960; Russell, 1970). The high value for K^+ exchange found by Russell and given in Table 2.9 includes mostly non-exchangeable K^+, suggested by Russell to be due to the introduction of K^+ ions into illitic-

TABLE 2.9 Exchange reactions involving suspended matter during mixing of river water and sea water.

	Rio Ameca[a]		Amazon[b]		Mean rivers[c]	
	Net charge (meq/100 g)	Fraction of CEC	Net charge (meq/100 g)	Fraction of CEC	Net charge (meq/100 g)	Fraction of CEC
H^+	—	—	+4.1	+0.17	+1.7	+0.06
Na^+	−8.0	−0.11	−9.3	−0.38	−12.1	−0.44
Mg^{2+}	−8.4	−0.11	−6.1	−0.25	−3.8	−0.14
K^+	−10.9	−0.14	−1.6	−0.07	−0.9	−0.03
Ca^{2+}	+29.5	+0.39	+12.9	+0.53	+15.1	+0.55
CEC^d	76 meq 100 g^{-1}		24.4 meq 100 g^{-1}		27.6 meq 100 g^{-1}	

− indicates uptake from sea water; + addition to sea water.
(a) Calculated from Russell (1970).
(b) Calculated from Sayles and Mangelsdorf (1977, 1979).
(c) This study, see text.
(d) Cation exchange capacity of the suspended load.

type layers in montmorillonites. However, existing data are too scarce to evaluate the importance of this process on a global scale. Thus, fluxes calculated here may be underestimated, especially for K^+ and Mg^{2+}.

   2.3.2.1.2 Biological uptake. The process of removal of calcium and silica by marine organisms in the water column is well known. Production of calcium carbonate and amorphous opaline silica by biological processes may be estimated from primary productivity and from the mean chemical composition of plankton. However, after death of the organisms and removal of the organic protective layer, the skeletons may undergo dissolution if they encounter water undersaturated with respect to their mineral composition. All ocean waters are undersaturated with respect to opaline silica (see Hurd, this volume, Chapter 6), whereas active dissolution of calcium carbonate occurs mainly in deep waters which are undersaturated with respect to both calcite and aragonite. Thus, silica and calcium, respectively, are regenerated in the water column during settling of siliceous and calcareous skeletons and, finally, only a small fraction of the initial production of these materials is deposited. An extended discussion of these processes can be found, for example, in the detailed review by Berger (1976). The net fluxes resulting from the biological processes recently were recalculated by Wollast (1981). We used the estimates of Wollast for the biogenic production of silica and calcium carbonate as equivalent to 12×10^{15} g Si y^{-1} and 3.5×10^{15} g Ca y^{-1}, respectively. Of these total amounts produced, only 375×10^{12} g Si y^{-1} and 600×10^{12} g Ca y^{-1} are deposited, which, in turn, represent the net removal of these elements from the water column by organisms.

   Besides SiO_2 and $CaCO_3$, which constitute the bulk of the skeletal material precipitated by marine organisms, impurities also are incorporated into the skeletons of organisms and removed from the water column. Of particular interest to this study is the work of Mackenzie *et al.* (1978) showing that the uptake of Al by diatoms was sufficient to account for the removal of dissolved aluminium carried by rivers to the ocean. Furthermore, calcite precipitated by marine organisms contains between 0 and 25 mole % magnesium carbonate. However, magnesium calcite skeletal materials principally accumulate in shallow waters. Shallow-water calcareous sediments contain about 5% $MgCO_3$ (Chave, 1954), whereas calcareous constituents of deep-water oozes contain about 0.5% $MgCO_3$ (Drever, 1974). Assuming that 20% and 80% of the Ca deposited annually are removed in shallow-water and deep-water calcareous sediments, respectively, we arrived at a biogenic Mg removal of 7×10^{12} g y^{-1}.

   2.3.2.1.3 Summary. The net fluxes owing to biological removal and cation exchange reactions are summarized in Table 2.10. Thus, we may calculate the net mass balance for dissolved species in the water column, which is reported in the last column of Table 2.10. The agreement for all the elements, except sodium and iron, is very satisfying and well within the uncertainties for the individual fluxes.

In the case of iron, it has been documented (Billen *et al.*, 1976; Boyle *et al.*, 1977) that "dissolved" iron carried by rivers is rapidly precipitated as hydroxides in the mixing zone with sea water and that the reduced dissolved iron released from anaerobic sediments also is rapidly precipitated under the oxic conditions prevailing in the water column. Iron is also precipitated as iron smectites, nontronite and hydrated iron oxides in the deep sea (Hein *et al.*, 1979; McMurtry and Yeh, 1981). Thus, it is likely that the remaining net flux of iron given in Table 2.10 is simply removed by these processes.

The imbalance in sodium is large; 65% of the river input is not accounted for in our mass balance calculations. However, there are major uncertainties in the estimation of the pore water flux of Na^+. An important sink for Na^+ on a geological time scale is the formation of evaporites. If the amount of unbalanced sodium is expressed in terms of halite deposition, it would correspond to 2.2×10^{14} g NaCl y^{-1} compared to a potential total depositional rate of 3.6×10^{14} g y^{-1}. There are no important NaCl deposits forming today; thus, one possibility is that Na^+ is accumulating in the ocean. If so, in 6×10^6 y at a rate of accumulation of 85×10^{12} g Na^+ y^{-1}, the average salinity of the ocean would increase only about $1^o/_{oo}$.

However, the Cl^- balance for the ocean indicates that it is likely that the major problem in the imbalance for Na^+ lies in the flux estimates for sediment pore waters, and perhaps, submarine weathering processes. Of the 115×10^{12} g Cl^- y^{-1} entering the ocean, 56% is trapped in pore waters balanced principally by Na^+. The excess Cl^-, 50×10^{12} g y^{-1}, is equivalent to 32×10^{12} g Na^+ y^{-1}. Thus, the imbalance in Na^+ from this calculation is $85 \times 10^{12} - 32 \times 10^{12} = 53 \times 10^{12}$ g Na y^{-1}, which cannot be accounted for by removal of NaCl. It is likely that this value is well within the uncertainties

TABLE 2.10 Net fluxes of dissolved species for the water column (in units of 10^{12} g y^{-1}).

	R^a	Exchange reactions[b]	Biological removal[c]	Net balance
Na	+128	−43	negligible	+85
Mg	+2	−7	−7	−12
Al	+1.9	0	−2	0
Si	+423	0	−375	+48
K	+9	−5	negligible	+4
Ca	+626	+47	−600	+73
Fe	+23.5	0	0	+23.5

(a) Calculated net mass balance for the water column from Table 2.8.
(b) Calculated from Table 2.9, assuming a total solid discharge of 15.5×10^{15} g y^{-1}, and a calculated CEC of 27.6 meq 100 g^{-1}.
(c) See text.

involved in the flux calculations for Na^+. There is a strong need for good analytical data on Na^+ concentrations in sediment pore waters and their distribution with depth.

2.3.2.2 Particulate materials—comparison of input and output

Comparison of the chemical and mineralogical composition of the input (river discharge and atmospheric dust) of particulate materials to the ocean with the composition of marine sediments is one way of assessing the major diagenetic processes operating in the marine system. To accomplish this comparison, we need to assess the global average composition of silicates in marine sediments.

2.3.2.2.1 Chemical composition. In Table 2.11 the average chemical composition is given for some typical marine sediments. It is possible from these data and other information to calculate an average chemical composition for marine sediments.

We can obtain an estimate of the relative proportions of shelf, hemipelagic, and pelagic sediments in the oceans by use of average sedimentation rates and areas for these regions. Assuming sedimentation rates of 0.5 (Worsley and Davies, 1979), 5 (e.g. Emery, 1969), and 12.5 (e.g. Gross, 1972) g cm^{-2} (1000 y)$^{-1}$ and areas of 268, 63 and 30×10^{16} cm^2, respectively, for pelagic, hemipelagic and shelf sediments, we arrive at relative proportions of 16, 38, and 46%. If we assume that all the sand and silt fraction of river suspended matter transported to the oceans is deposited in the coastal zone (an overestimate), and compare the Si and Al content of the river suspended load (Table 2.1) with concentrations of these elements in marine sediments (Table 2.11), it appears that 7% of river suspended solids as sand and silt is deposited in nearshore marine areas. Subtracting this value from the relative mass proportion of shelf sediments of 46%, leads us to the conclusion that

TABLE 2.11   Average chemical composition of marine sediments (wt. %).

	Nearshore mud[a]	Nearshore sand[a]	Hemipelagic blue and green muds[b]	Pelagic red clay[c]
SiO_2	55.02	78.50	57.05	53.93
Al_2O_3	16.61	6.12	17.22	17.46
Fe_2O_3	7.26	4.69	7.63	9.03
MgO	2.19	0.89	2.17	4.56
CaO	1.57	3.90	2.04	1.56
Na_2O	2.76	1.18	1.05	1.23
K_2O	2.32	1.14	2.25	3.65

(a) From Hirst (1962). Na content seems abnormally high.
(b) From Clarke (1924).
(c) From El Wakeel and Riley (1961).

39 % of the total mass of marine sediments is represented by nearshore mud. It now is possible to calculate an average composition of marine sediments by multiplying each composition in Table 2.11 by its corresponding mass fraction. The results of this calculation are given in Table 2.12, column 1.

The average composition of marine sediments also can be calculated from the total input of particulate and dissolved materials to the sea. Because we are concerned mainly with the cycle of silicates, we have subtracted from the total annual flux of material to the sea the amount of calcium carbonate accumulated in marine sediments, as given in Section 2.3.2.1.2. Using the average composition of limestone given by Clarke (1924) to represent elements associated with carbonate rocks, we then can adjust the remaining fluxes accordingly and arrive at an estimate of the composition of the silicate portion of marine sediments (Table 2.12, column 2).

The two calculations leading to estimates of the average composition of carbonate-free marine sediments agree remarkably well, except for iron. We now may compare these two estimations to the average composition of river suspended materials (Table 2.1), and calculate the chemical changes between terrigenous silicate materials and marine, assuming Al_2O_3 is conserved. The results are given in columns 3 and 4 of Table 2.12. It appears that terrigenous solids after they enter the ocean gain Si, Mg, Na, and K and lose Ca in their chemical evolution to marine sediments. Iron probably is added to the marine sediment mass by reactions involving submarine basalt and sea water (Hein et al., 1979; McMurtry and Yeh, 1981). The substantial difference for Fe_2O_3 between columns 4 and 5, Table 2.12, suggests we may have underestimated the submarine flux of Fe.

2.3.2.2.2 *Mineralogical composition.* The chemical changes described above should be reflected in the mineralogy of sediments. Unfortunately,

TABLE 2.12 Average composition of carbonate-free marine sediments (wt. %) and chemical evolution of terrigenous materials to marine sediments.

| | Average composition | | Compositional change (3) | |
	(1)	(2)	from column (1)	from column (2)
SiO_2	57.16	57.36	+ 1.57	+ 2.59
Al_2O_3	16.26	16.00	—	—
Fe_2O_3	7.56	6.37	+ 1.42	0.19
MgO	2.56	2.35	+ 0.60	+ 0.41
CaO	1.89	1.86	− 0.87	− 0.87
Na_2O	1.75	1.88	+ 0.97	+ 1.14
K_2O	2.48	2.52	+ 0.31	+ 0.40

(1) Computed from the composition of average marine sediments.
(2) Computed from total input of materials to the marine system corrected for carbonate deposition.
(3) See text.

existing data concerned with the quantitative mineralogical composition of sediments are scarce, except for pelagic clays which have been more thoroughly investigated. The mineralogical composition of the clay fraction of marine sediments, which represents 40–80 % of the sediment mineralogy, was reviewed recently by Windom (1976), and is summarized in Table 2.13. We have computed from these data a weighted average (Table 2.13) by taking into account the surface area of each ocean. Windom also shows that quartz contents of marine sediments vary from 4 to 13 %, with an average of 8.5 %, and that the ratio of feldspar to quartz is nearly one.

TABLE 2.13   Composition of the clay fraction of marine sediments (wt. %).[a]

	Kaolinite	Chlorite	Montmorillonite	Illite
North Atlantic	20	10	16	56
South Atlantic	17	11	26	47
North Pacific	8	18	35	40
South Pacific	8	13	53	26
Indian Ocean	16	10	47	30
Weighted Average[b]	12.5	13	38.4	37.3

(a) After Windom (1976).
(b) This paper.

To obtain a more quantitative picture of possible reactions involving silicates in the marine system, we attempted to apply normative calculations as a means of evaluating mineralogical distributions. However, determination of the normative minerals in sedimentary rocks is a difficult task because of the wide range of mineralogical composition encountered in nature. We chose a number of end-member compositions to serve as normative minerals; the compositions of the clay minerals selected are given in Table 2.14. These normative mineral compositions do not necessarily represent the exact compositions of minerals found in nature but enable us to compare sedimentary patterns of a diverse nature and to make a tentative mass balance among sedimentary materials. The end-member compositions provide a clue as to the relative abundances of particular minerals and reflect the bulk chemistry of the materials. For example, as the illite component of a mixed layer smectite/illite increases in amount in a sediment, the potassium content may increase and the Si/Al ratio of the sediment decrease. This compositional change will be reflected in a calculated increase in the amount of normative illite.

It should be pointed out that the chemical composition alone is insufficient to carry out normative calculations involving several phases of complex composition. Therefore, major mineralogical trends in various sedimentary rocks also were taken into account. Besides the classical group

TABLE 2.14   Normative mineral compositions of clay minerals.

Na montmorillonite[a]	$Na_{0.4}Mg_{0.34}Fe_{0.21}Al_{1.67}Si_{3.83}O_{10}(OH)_2$
Ca montmorillonite[b]	$Ca_{0.2}Mg_{0.34}Fe_{0.21}Al_{1.67}Si_{3.83}O_{10}(OH)_2$
Illite[a]	$K_{0.8}Mg_{0.34}Fe_{0.21}Al_{2.07}Si_{3.43}O_{10}(OH)_2$
Chlorite[b]	$Mg_2Fe_3Al_2Si_3O_{10}(OH)_8$
Kaolinite	$Al_2Si_2O_5(OH)_4$
Vermiculite[c]	$(Mg, Ca)_{0.5}(Mg_{2.5}Fe_{0.33}Al_{1.67}Si_{2.5})O_{10}(OH)_2$
Glauconite[d]	$Mg_{0.29}Fe_{1.05}Al_{1.27}Si_{3.53}O_{10}(OH)_2$
Sepiolite	$Mg_2Si_3O_8 \cdot 4H_2O$

(a) From Tardy and Garrels (1974).
(b) From Garrels and Mackenzie (1971).
(c) High magnesium vermiculite based on chemical composition given in Deer, Howie, and Zussman (1962).
(d) Computed from the average chemical composition of "young marine" glauconite given by Calvert (1976).

of minerals, montmorillonite, illite, kaolinite, chlorite, we have also introduced into the calculations other minor minerals commonly reported from marine sediments like zeolites, attapulgite—sepiolite, vermiculite, and glauconite (Windom, 1976; Calvert, 1976). In these calculations, zeolites were not distinguished from feldspars because they differ only by the addition of water. The results of the calculations are shown in Table 2.15.

For the major clay minerals, analysis of Table 2.15 provides some interesting information. First, the distribution of montmorillonite, illite, kaolinite and chlorite in blue and green muds is very similar to the

TABLE 2.15   Normative calculation for the mineralogical composition of various sediments (wt. %).

	River suspended sediment	Nearshore sand	Blue and green muds	Red clay
Na montmorillonite	15	8	32	24
Ca montmorillonite	15	—	—	—
Vermiculite	1	—	1	3
Glauconite	—	—	2	—
Illite	15	4	16	22
Kaolinite	8	2	8.5	6
Chlorite	4	1	4.5	8
Sepiolite	—	—	—	3
Na Feldspar	3	8	2.5	4
K Feldspar	6	4	5.0	8
Ca Feldspar	3	—	—	—
Quartz	23	62	20	12
Goethite	4	4	4	4
Calcite	3	7	3	3

distribution of these minerals in the river suspended load, with the exception that Ca montmorillonite has been transformed into Na montmorillonite. Authigenic occurrences of recent glauconite in organic-rich, nearshore sediments and blue and green muds now are well documented (Calvert, 1976), and this mineral appears in the normative calculation for blue and green muds.

Second, comparison between river suspended materials and pelagic clays is informative; the major components of the clay fraction are given in Table 2.16. Two different processes may explain the differences observed in Table 2.16. First, only the finest fraction of clay minerals which have undergone

TABLE 2.16   Comparison between river suspended sediment and pelagic clay for the components of the clay fraction (wt. %).

	River suspended load	Pelagic clays
Smectite	53	41
Illite	26	36
Kaolinite	15	10
Chlorite	6	13

minor flocculation probably are transported to the deep sea. However, it is not easy to elucidate factors influencing the clay mineral dispersal pattern because these patterns are influenced substantially by various continental detrital sources (Griffin and Goldberg, 1963; Mowatt and Naid, 1979; Windom, 1976). Furthermore, there is little difference between the bulk mineralogical composition of the clay fraction deposited in nearshore and hemipelagic realms and river suspended solids, as would be expected if significant sorting of clay minerals occurred. Thus, the differences observed in Table 2.16 reflect most probably diagenetic processes.

The apparent increase of the content of illite and chlorite in pelagic clays with respect to river suspended clays is in good agreement with the increase of K_2O and MgO shown by the average chemical analysis of these sediments. It is also in agreement with the uptake of K^+ and Mg^{2+} observed in pore waters. Furthermore, the increase of normative illite and chlorite may be exactly balanced by the decrease of montmorillonite and kaolinite according to the following reactions:

$$Na_{0.40}Mg_{0.34}Fe_{0.21}Al_{1.67}Si_{3.83}O_{10}(OH)_2 + 0.80K^+ +$$
$$0.20Al_2Si_2O_5(OH)_4 + 0.40HCO_3^- + H_2O$$
$$= K_{0.8}Mg_{0.34}Fe_{0.21}Al_{2.07}Si_{3.43}O_{10}(OH)_2 + 0.40Na^+ +$$
$$0.40CO_2 + 0.80H_4SiO_4 \tag{12}$$

and

$$Al_2Si_2O_5(OH)_4 + 5Mg^{2+} + 10HCO_3^- + H_4SiO_4$$
$$= Mg_5Al_2Si_3O_{10}(OH)_8 + 10CO_2 + 5H_2O \tag{13}$$

These reactions are analogous to the "reverse weathering" reactions of Mackenzie and Garrels (1966a,b). Using the observed decrease of 0.029 mol Na montmorillonite/100 g clay and the observed increase of 0.012 mol chlorite/100 g clay, calculation of the corresponding increase in illite and decrease in kaolinite according to the above equations are, respectively, $+0.029$ mol 100 g^{-1} clay and -0.018 mol 100 g^{-1} clay, in good agreement with the $+0.028$ and -0.017 mol 100 g^{-1} clay observed. It should be emphasized that the above reactions involve normative minerals; the actual phases involved may include minerals like vermiculite and amorphous aluminosilicates.

The increase of Na_2O in marine sediments is not reflected by mineralogical changes in the major constituents of the clay fraction. The transformation of Na montmorillonite to illite should release Na^+, and could account for the release of Na^+ to pore waters observed by Sayles (1979). On the other hand, the occurrence of phillipsite and authigenic feldspars in marine sediments is common, especially in regions of volcanogenic sediments (Elderfield, 1976), and could explain the relative enrichment of Na_2O. Finally, the net effect of the two reactions discussed above is a release of Si to pore waters. The concentration of dissolved silica in pore waters is generally greater than the solubility of quartz and SiO_2 overgrowths on this mineral have been observed both in laboratory experiments (Mackenzie and Gees, 1971) and in sediment cores of the Deep Sea Drilling Project. Formation of authigenic feldspars is another possible mechanism for silica precipitation from pore waters.

The above reactions, based on Mg^{2+} and K^+ pore water fluxes, could conceivably return to the atmosphere as much as 80% of the 10.4 × 10^{12} moles CO_2 y^{-1} used in the chemical weathering of silicate minerals. However, the CO_2 demand for weathering of Na- and Ca-bearing silicates is about equal to that for Mg- and K-bearing phases, suggesting that the CO_2 flux calculated from equations (12) and (13) may be overestimated. Whatever the case, the calcium released from the weathering of calcium silicates is deposited in the ocean as calcium carbonate. The CO_2 utilized in the weathering process may be returned by late diagenetic and metamorphic reactions like:

$$CaCO_3 + SiO_2 = CaSiO_3 + CO_2. \tag{14}$$

It appears that metamorphic (late diagenetic) and early diagenetic reactions (like (12) and (13)) may be of about equal importance in returning CO_2 used in silicate weathering to the atmosphere, about 5 × 10^{12} moles CO_2 y^{-1} each. Because the intensity of metamorphism is episodic, the imbalance of respiration plus oxidation over photosynthesis in the ocean, resulting in the release of CO_2 from the ocean, may temporarily help to prevent a drain on atmospheric CO_2 by silicate weathering (Garrels and Mackenzie, 1972).

2.4   SILICA IN THE ROCK CYCLE

Discussion of the cycling behaviour of silica in the oceans and relatively young sediments provides a basis for understanding the circulation of this element in the rock cycle. Accumulation of silica in marine sediments is but an initial stage in a series of processes that affect silica during diagenesis and metamorphism. In this section the fate of silica in the rock cycle is discussed, as well as factors that may influence silica fluxes in the global cycle.

2.4.1   Comparison of Present-day and Geologic Cycles

The fluxes, estimated in this paper, associated with today's global cycle of silica are shown in Fig. 2.2. For comparison a model of the circulation of

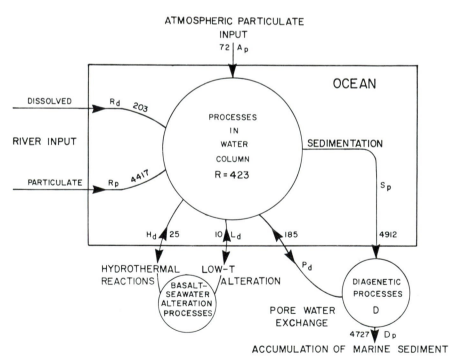

FIG. 2.2   Global cycles of silica. Fluxes are in units of 10^{12} g Si y^{-1} (see Fig. 2.1 for additional explanation).

silicon during geologic time in the sedimentary cycle is illustrated in Fig. 2.3. Construction of this model is based on the concept that the sedimentary rock cycle is in steady state. Justification for this assumption and detailed discussion of element circulation in the sedimentary rock cycle are given in Garrels and Mackenzie (1972), and Mackenzie (1975).

Two major rock reservoirs and the oceanic reservoir of silicon are represented in the silicon cycling model of Fig. 2.3. New rocks are more susceptible to post-depositional change and erosion than old rocks; therein lies the major reason for dividing the total sedimentary mass of 25 000 × 10^{20} g into two parts. The major rock reservoirs are further subdivided into sub-reservoirs representing the masses of silicon in the major rock types of shale, sandstone, limestone, chert, and evaporite. The recent chemical compositional data of Ronov (1980) and the relative abundances of lithologies in the "new" and "old" rock reservoirs as given by Garrels and Mackenzie (1971, 1972) are used to estimate silicon masses within the various lithologies. The silicon fluxes among reservoirs are revised values of Garrels and Mackenzie's (1972) original estimates.

Comparison of the present-day silica cycle with the sedimentary rock cycle of silicon points to a number of interesting relations.

(1) The total flux of silicon from weathering to the ocean to sediments today is three to four times greater than the steady-state flux during geologic time. Today's flux is dominated by silicon in the suspended load of rivers; the ratio of silicon in the suspended load to silicon in the dissolved load today is 22:1, whereas in the steady-state sedimentary cycle model it is 6:1.

This difference reflects mainly the fact that today's ratio of total suspended

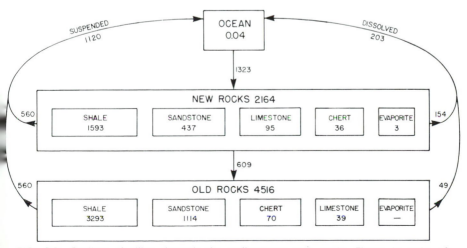

FIG. 2.3 Cycling of silica through the sedimentary rock cycle. Fluxes in units of 10^{12} g Si y^{-1}; masses in units of 10^{20} g Si (modified from Garrels and Mackenzie, 1972).

to dissolved load of about 4:1 (Meybeck, 1977) is high compared to the ratio of 1:1 to 2:1 for the geologic past (Garrels and Mackenzie, 1972). Most riverine silicon is transported in the suspended load. The relatively high mechanical to dissolved load ratio of today's rivers is due to man's deforestation and agricultural activities (Judson, 1968) modified by the facts that today's continents are unusually high and land area is relatively large compared to the long-term geological average for these factors. Both latter factors lead to increased mechanical and chemical denudation rates but the mechanical load increases more rapidly with increasing mean continental elevation and land area than the dissolved load (cf. Pigott, 1981; Garrels and Mackenzie, 1971; Meybeck, 1979).

(2) The steady-state sedimentary cycle of Fig. 2.3 is closed with respect to inputs or outputs of materials stemming from reactions between sea-floor basalts and seawater. It can be seen in Fig. 2.2 that of the total dissolved silica entering the oceans today about 8% is a product of hydrothermal and low-temperature reactions involving basalts and sea water. The residence time of this silica, defined as *total silica in sedimentary rocks/input of dissolved silica from basalt–seawater reactions*, is about 40 times greater than the total residence time of silica in sedimentary rocks (Fig. 2.3; 6680×10^{20} g Si/1323 $\times 10^{12}$ g Si y $= 500 \times 10^6$ y). Thus, in terms of the total amount of silicon circulating in the sedimentary rock cycle, recognition of the component derived from basalt-seawater reactions is difficult.

(3) With respect to the circulation of dissolved silica in the rock cycle today, the residence time of dissolved silica in the oceans related to the flux of silica from basalt-seawater reactions $(0.04 \times 10^{20}$ g/35×10^{12} g y^{-1} $= 117\,000$ y) is about six times that with respect to river input of silica derived from continental weathering $(0.04 \times 10^{20}$ g/203×10^{12} g y^{-1} $= 20\,000$ y). At the current rate of crust production at ridges of 3 km^2 sea-floor y^{-1}, the entire oceanic crust would be replaced in 100×10^6 y. In this period of time, if the oceanic reservoir of silica were in steady state, 34×10^{20} g Si produced by basalt–seawater reactions could be deposited on the pelagic sea-floor as biogenic amorphous silica or as silicate minerals such as authigenic Fe-rich smectite and zeolites (Hein et al., 1979). This amount of silica represents a deposition rate of 0.026 g SiO_2/cm^2 1000 y over the entire area of pelagic seafloor. Accumulation rates of opaline silica in deep-sea sediments of the Pacific Ocean vary from less than 0.005 to much greater than 0.15 g/cm^2 1000 y (Bostrom et al., 1973). Holocene silica accumulation rates in the Pacific and Indian Oceans vary from maximum values of greater than 3 g/cm^2 1000 y to less than 0.05 g/cm^2 1000 y, and generally are highest in the equatorial and high latitude regions of high productivity of siliceous organisms (Lisitzin, 1972). Figure 2.2 shows that 423 $\times 10^{12}$ g Si y^{-1} are deposited on the sea-floor principally as biogenic opaline

silica, representing a rate of accumulation of silica of $0.24 \, g/cm^2 \, 1000 \, y$ over the entire sea-floor.

(4) The fate of silica derived from basalt–seawater alteration processes and sedimented on the deep-sea floor is difficult to assess. That portion which is not subducted but is incorporated in sediments scraped off descending oceanic plates and plated onto continents, represents a chemical interchange of silica (and other materials derived from basalt–seawater reactions) between the shallow crustal reservoirs of Fig. 2.3 and the mantle. The problem is that the magnitude of this silica flux even for the present is not well known.

Certainly some portion of biogenic siliceous oozes deposited in the deep sea is added to continents, as evidenced by outcrops of these deposits on oceanic islands and in continental settings. Some reddish-brown cherts and silicified mudstones found in geosynclinal sequences exposed on continents probably represent sea-floor offscrapes. However, the amount of silica in these deposits derived from basalt–seawater reactions is not known. An order of magnitude estimate of the flux of dissolved silica from the mantle to the shallow crustal (exogenic) sedimentary cycle of Fig. 2.3 is that involved in basalt–seawater reactions today, $35 \times 10^{12} \, g \, Si \, y^{-1}$.

(5) The relative importance of riverine and sea-floor sources of dissolved silica to the ocean probably varied during geologic time. Today's ratio of river-derived to sea-floor basalt-derived dissolved silica is 6:1.

In a series of excellent articles, Meybeck and Martin (see references) have defined the major factors influencing the dissolved and particulate transport rates of the world's rivers on a global scale. Meybeck (e.g. 1979) has shown that river run-off is the key factor affecting total dissolved material transport by rivers to the ocean; of lesser importance are the factors of temperature, lithology, and relief. Dunne (1978) has demonstrated that chemical denudation rates of silicate rocks in 43 Kenyan river catchments increase with increasing mean annual run-off from the river basins. Also, for these basins, as mean annual run-off increases, the proportion of dissolved silica in the total dissolved load increases. Thus, it seems most likely that on a global scale, variations in the rate of transport of dissolved silica to the oceans will be determined principally by variations in river water run-off from the land. Variations in river discharge from the continents to the ocean, in turn, are related principally to the latitudinal positions (and hence climate and vegetation) and total surface area of the continents, perhaps modified slightly by mean continental elevation (Garrels and Mackenzie, 1971; Meybeck, 1979).

At our present stage of knowledge of basalt–seawater reactions and mass transfer, changes in the rate of sea-floor spreading presumably would result in changes in element fluxes (Wolery and Sleep, 1981). For dissolved silica, it is likely that an increase in the rate of sea-floor spreading, leading to an increase in the rate of production of oceanic crust and in the average temperature of hydrothermal basalt–seawater reactions (Wolery, 1978), would give rise to an increase in the flux of dissolved silica to the oceans.

Sea level has varied during the last 600×10^6 y of earth's history (Vail *et al.*, 1977). The long-term variations in sea level are most likely related to changes in sea-floor spreading rates and ridge volumes (Hays and Pitman, 1973). Fast sea-floor spreading rates and large ridge volumes give rise to high sea levels; slow rates and small volumes result in low sea levels. Other causes affect interpretation of the sea-level curve of Vail *et al.* (glaciations, basin subsidence, and sedimentation rates); however, it is most likely the case that sea level was indeed high and ridge volumes large during the Cretaceous and late Cambrian to Ordovician relative to the low sea-level stands and small ridge volumes of most of the latter part of the Tertiary, the late Carboniferous to early Jurassic, and the late Proterozoic to early Cambrian.

Based on the above relations and discussion, we conclude that the fluxes of dissolved silica to the ocean via rivers and basalt–seawater reactions varied during the last 600×10^6 years of earth's history. Times of high sea-level, inudation of the continents (small land area, low run-off), rapid sea-floor spreading, and high ridge volume were times of maximum input of dissolved silica to the oceans from basalt–seawater reactions relative to its continental flux. During global low stands, the continental flux increased in importance relative to the flux involved with basalt–seawater reactions. The latitudinal positions of continents probably exerted a modifying effect on the continental flux of silica. For example, during the early Cambrian, continents of that time were widely distributed about the equatorial zone (Ziegler *et al.*, 1979). Such a distribution may have resulted in a higher river run-off and silica flux to the oceans than would have obtained if the continents were distributed more uniformly about the earth's surface.

2.4.2 Diagenetic Processes—Their Importance in the Global Silica Cycle

Silicon is transferred from the weathering environment on continents to oceans as quartz or in the clay minerals of the suspended load of rivers. In addition, because of weathering and diagenetic processes, which release silica from silicate minerals, silica is an important dissolved constituent of rivers. Today, most of the dissolved silica entering the oceans is removed as opaline silica in the tests of diatoms, radiolarians, and silicoflagellates. During diagenesis this opaline silica and some detrital silicates are transformed to other phases. In some of these transformation processes, dissolved silica is transferred from one rock type to another.

Figure 2.4 illustrates the complex web of major diagenetic reactions that silica and silicate minerals undergo during burial, as these materials enter environments of increasing temperature and pressure, encounter solutions of different compositions, and age. The diagram is not meant to be complete, as the actual diagenetic history of silicates is more complex than shown, but it does provide a means of presentation of the major pathways of movement of silica during burial and uplift of silicate minerals. The following discussion is keyed to Fig. 2.4.

Weathering of silicate minerals in igneous and sedimentary rocks produces silica in solution and clay minerals. Sedimentary rocks include normal sediments such as sandstones and shales as well as metamorphic rocks that still retain sedimentary chemical compositions. Igneous rocks include primary rocks condensed from molten magma as well as sedimentary rocks that have been reconstituted completely and no longer chemically resemble sedimentary rocks (Garrels and Mackenzie, 1971).

Silicon in detrital silicates is transported to the oceans as suspended load of rivers and sediments on the sea-floor. Dissolved silica from continental weathering and submarine basalt–seawater reactions is deposited principally as amorphous silica but some amount may precipitate from seawater as authigenic silicates such as zeolites, smectites, sepiolite, and palygorskite.

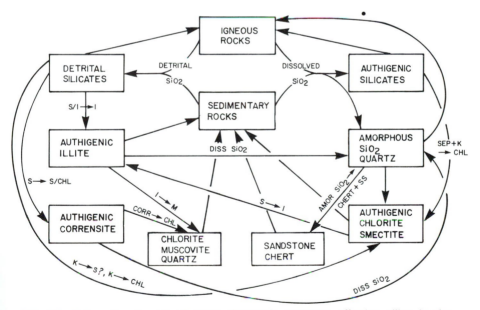

FIG. 2.4 Schematic diagram of major diagenetic processes affecting silica in the sedimentary rock cycle. S/I, I, S/Chl, S, K, Corr, M, Amor SiO$_2$, SS, Sep and Chl refer to the materials smectite/illite, illite, smectite/chlorite, smectite, kaolinite, corrensite, mica, amorphous silica, sandstone, sepiolite, and chlorite, respectively.

This precipitation process occurs at the sediment water interface and during shallow burial. The metastable mineral assemblage of both detrital and authigenic phases found in recent marine sediments is then subjected to a complicated set of changes that eventually convert many of the minerals of recent sediments to other phases (Garrels and Mackenzie, 1974). New rocks (principally of Cenozoic and Mesozoic age) are characterized by an abundance of silicate phases of relatively high silica content; in general, with time and increasing temperature, these phases by reaction with aluminous phases such as kaolinite or gibbsite are converted to silicate minerals found in old rocks (principally of Paleozoic and Precambrian age; Fig. 2.3).

One of the best documented diagenetic pathways of Fig. 2.4 is that involving the conversion of mixed layer smectite/illite to illite in shales (e.g. Perry and Hower, 1970; Hoffman and Hower, 1979; Boles and Franks, 1979). The illitization reaction using "average" montmorillonite and "end-member" illite (Garrels and Mackenzie, 1974; Tardy and Garrels, 1974) is:

$$K_{0.40}[(Mg_{0.34}Fe^{3+}_{0.17}Fe^{2+}_{0.04}Al_{1.50})(Al_{0.17}Si_{3.83})O_{10}(OH)_2]$$
$$+ 0.4KAlSi_3O_8$$
$$= K_{0.80}[(Mg_{0.34}Fe^{3+}_{0.17}Fe^{2+}_{0.04}Al_{1.50})(Al_{0.57}Si_{3.83}O_{10}(OH)_2]$$
$$+ 1.6H_4SiO_4 \tag{15}$$

average montmorillonite + K-spar = end-member illite + dissolved silica. The dissolved silica released during this reaction precipitates in the shale as quartz (Yeh and Savin, 1977) or is transported out of the shale to intercalated sandstones where it may precipitate as silica overgrowths on detrital quartz grains or as interstitial silicate cement. The precipitation of quartz may drive the above reaction to the right during burial diagenesis (Tardy and Garrels, 1974); however, an increase in temperature during diagenesis increases the rate of conversion of smectite layers to illite (Hower et al., 1976; Eberl and Hower, 1976), provided sediment pore waters contain less dissolved silica than that required for the stable coexistence of montmorillonite–K-spar.

Boles and Franks (1979) have demonstrated that the smectite-illite transformation is more complex than shown in reaction (15) because the actual compositional complexity of mixed layer smectite/illite and its transformation to illite may require release of dissolved constituents other than silica (e.g. Na^+, Ca^{2+}, Fe^{2+}, Mg^{2+}). These constituents also may leave shale systems via water flow and enter sandstones. The quantitative importance of this mass transfer from shale to sandstone is such that during diagenesis of a shale sequence 100 m thick a sandstone body 17 m thick with an initial porosity of 25% could be completely cemented by silica and carbonate cements (Boles and Franks, 1979).

The ultimate end-product in the diagenesis of smectite is dioctahedral mica (Maxwell and Hower, 1967; Muffler and White, 1969). This mineral along with chlorite and quartz is the characteristic silicate mineral assemblage found in shales of old rocks. The chlorite is derived via a number of reaction pathways shown in Fig. 2.4. This assemblage plus chert and recycled chert in sandstones represents the final product in the stabilization of silicate minerals when silicate assemblages in new rocks are converted to silicate assemblages in old rocks. The overall transformation process involves release of dissolved silica to pore waters; some of this dissolved silica, along with that derived from chemical weathering of minerals, enters surface waters during uplift of sedimentary rocks (Fig. 2.3). Thus, dissolved silica originally obtained from surface processes, deposited in the ocean and buried is returned to the surface environment. Diagenetic processes play a significant role in the circulation of silica through the sedimentary rock cycle.

2.5 CONCLUSIONS

Study of the global cycle of an element and its interactions with other substances is complex. Silicon is one of the better known elements in terms of reservoir masses, fluxes, and processes affecting it. However, even in the case of silicon, it is difficult to arrive at an unique picture of its present-day circulation in the exogenic cycle. In terms of the cycling behaviour of silicon on a geologic time scale and the role of deep-crustal and mantle processes in its circulation, we are only beginning to come to grips with the problems. It is hoped that this article has demonstrated the myriad of facts, estimates, and considerations that need attention in description of the global cycle of an element. Full description requires not only evaluation of the effect of surface processes on element circulation but also processes and mass transfer involved in diagenesis and metamorphism.

ACKNOWLEDGEMENTS

We thank Virginia Paterson for drafting the figures. Research supported by NSF Grant EAR 76-12279.

REFERENCES

Alekin, O. A. and Brazhnikova, L. V. (1968). *Ass. Int. Hydrol. Sci. Pub.* **78**, 35–41.
Alekin, O. A. and Brazhnikova, L. V. (1962). *Dokl. Acad. Nauk SSSR*, **146**, 203–206 (in Russian).
Alekin, O. A. and Brazhnikova, L. V. (1960). *Gidrochim. Mat.* **32**, 12–34 (in Russian).
Berger, W. H. (1976). Biogenous deep sea sediments. *In* "Chemical Oceanography".

2nd Ed. Vol. 5 (J. P. Riley and R. Chester, eds.). Academic Press, London and New York.

Billen, G., Smitz, J., Somville, M., and Wollast, R. (1976). Degradation de la matiere organique et processus d'oxydo-reduction dans l'estuaire de l'Escaut. *In* "Projet Mer., v. 10, L'Estuaire de L'Escaut" (J. C. J. Nihoul and R. Wollast, eds.). Serv. Prem. Ministre. Prog. Politique Scient., Bruxelles, Belgium.

Bischoff, J. L. and Sayles, F. L. (1972). *J. Sed. Pet.*, **42**, 711–724.

Blatt, H., Middleton, G., and Murray, R. (1972). "Origin of Sedimentary Rocks". Prentice Hall Inc., Englewood Cliffs, N.J.

Boles, J. R. and Franks, S. G. (1979). *J. Sed. Pet.*, **49**, 55–70.

Boström, K., Kraemer, T., and Gartner, S. (1973). *Chem. Geol.*, **11**, 123–148.

Boyles, E. A., Edmond, J. M., and Sholkovitz, E. R. (1977). *Geochim. Cosmochim. Acta*, **41**, 1313–1324.

Buat-Menard, P. and Chesselet, R. (1979). *Earth and Planet. Sci. Lett.*, **42**, 399–411.

Calvert, S. E. (1976). Mineralogy and geochemistry of near-shore sediments. *In* "Chemical Oceanography". 2nd Ed. Vol. 6 (J. P. Riley and R. Chester, eds.). Academic Press, London and New York.

Carroll, D. and Starkey, H. C. (1960). Effect of sea-water on clay minerals. *In* "Clays and Clay Minerals, 7th Nat. Conf." (A. Swinford, ed.). Pergamon Press, Oxford.

Chave, K. E. (1954). *J. Geol.*, **62**, 587–599.

Clarke, F. W. (1924). "Data on Geochemistry". US Geol. Sur. Bull. **770**.

Deer, W. A., Howie, R. A., and Zussman, J. (1962). "Rock-Forming Minerals". Vol. 3. J. Wiley, New York.

Delany, A. C., Parkin, D. W., Griffin, J. J., Goldberg, E. D., and Reiman, B. E. F. (1967). *Geochim. Cosmochim. Acta*, **31**, 885–909.

Drever, J. I. (1974). The magnesium problem. *In* "The Sea". Vol. 5 (E. D. Goldberg, ed.). Wiley-Interscience, New York.

Dunne, T. (1978). *Nature*, **274**, 244–246.

Eberl, D. D. and Hower, J. (1976). *G.S. A. Bull.* **87**, 1326–1330.

Edmond, J. M., Measures, C., McDuff, R. E., Cham, L. H., Collier, R., Grant, B., Gordon, L. I., and Corliss, J. B. (1979). *Earth Planet. Sci. Lett.*, **46**, 1–18.

Elderfield, H. (1976). Hydrogenous material in marine sediments. *In* "Chemical Oceanography". 2nd Ed. Vol. 5 (J. P. Riley and R. Chester, eds.). Academic Press, London and New York.

El Wakeel, S. K. and Riley, J. P. (1961). *Geochim. Cosmochim. Acta*, **25**, 110–146.

Emery, K. O. (1969). *Oil Gas J.*, **67**, 231–243.

Fanning, K. A. and Pilson, M. E. (1974). *J. Geophys. Res.*, **79**, 1293–1297.

Garrels, R. M. and Mackenzie, F. T. (1974). Chemical history of the oceans deduced from post-depositional changes in sedimentary rocks. *In* "Studies in Paleo-Oceanography" (W. Hay, ed.). *S. E. P. M. Spec. Pub.* **20**, 193–204.

Garrels, R. M. and Mackenzie, F. T. (1972). *Mar. Chem.*, **1**, 27–41.

Garrels, R. M. and Mackenzie, F. T. (1971). "Evolution of Sedimentary Rocks". Norton and Co., New York.

Gibbs, R. J. (1967a). *Geol. Soc. Am. Bull.*, **78**, 1203–1232.

Gibbs, R. J. (1967b). *Science*, **156**, 1734–1737.

Griffin, G. M. (1962). *Geol. Soc. Am. Bull.*, **73**, 737–768.

Griffin, J. J. and Goldberg, E. D. (1963). Clay mineral distribution in the Pacific Ocean. *In* "The Sea". Vol. 3 (M. N. Hill, ed.). Wiley-Interscience, New York.

Gross, M. G. (1972). "Oceanography". Prentice-Hall, New Jersey.

Hart, R. (1973). *Can. J. Earth. Sci.*, **10**, 799–816.

Heath, G. R. (1974). Dissolved silica and deep-sea sediments. *In* "Studies in Paleo-Oceanography" (W. W. Hay, ed.). *S.E.P.M. Spec. Pub.* **20**, Tulsa, Oklahoma.

Hein, J. R., Ross, C. R., and Alexander, E. (1979). Mineralogy and diagenesis of surface sediments from domes A, B, and C. *In* "Mar. Geol. Oceanog. Pac. Mangan. Nodule Prov." (J. L. Bischoff and K. Z. Piser, eds.). Plenum, New York.

Hirst, D. H. (1962). *Geochim. Cosmochim. Acta.*, **26**, 309–334.

Hoffman, J. and Hower, J. (1979). Clay mineral assemblages as low grade metamorphic geothermometers: Application to the thrust faulted disturbed belt of Montana USA. *In* "Aspects of Diagenesis" (P. A. Scholle and P. R. Schluger, eds.). *S.E.P.M. Spec. Pub.* **26**, 55–79.

Holland, H. D. (1978). "The Chemistry of the Atmosphere and Oceans". Wiley-Interscience, N.Y.

Horn, M. R. and Adams, J. A. (1966). *Geochim. Cosmochim. Acta*, **30**, 279–297.

Hower, J., Eslinger, E., However, M. E., and Perry, E. A. (1976). *G. S. A. Bull.*, **87**, 725–737.

Hurd, D. C. (1973). *Geochim. Cosmochim. Acta.*, **37**, 2257–2282.

Judson, S. (1968). *Am. J. Sci.*, **56**, 356–374.

Kennedy, V. C. (1965). *U.S. Geol. Sur. Prof. Paper* **433**-D, 25 pp.

Lisitzin, A. P. (1972). "Sedimentation in the World Ocean". *S.E.P.M. Spec. Pub.* **17**.

Livingstone, D. A. (1963). Chemical composition of rivers and lakes. *In* "Data of Geochemistry". 6th Ed. (M. Fleischer, ed.). *U.S. Geol. Surv. Prof. Paper* 440 G.

Loughnan, F. C. (1969). "Chemical weathering of the Silicate Minerals". Elsevier, New York.

Mackenzie, F. T. (1975). Sedimentary cycling and evolution of sea water. *In* "Chemical Oceanography." Vol. 1 (J. P. Riley and G. Skirrow, eds.). Academic Press, London and New York.

Mackenzie, F. T. and Garrels, R. M. (1966a). *Am. J. Sci.*, **264**, 507–525.

Mackenzie, F. T. and Garrels, R. M. (1966b). *J. Sed. Pet.*, **36**, 1075–1084.

Mackenzie, F. T. and Gees, R. (1971). *Science*, **173**, 533–535.

Mackenzie, F. T., Stoffyn, M., and Wollast, R. (1978). *Science*, **199**, 680–682.

Mackenzie, F. T. and Wollast, R. (1977). Thermodynamic and kinetic controls of global chemical cycles of the elements. *In* "Global Chemical Cycles and their Alteration by Man" (W. Stumm, ed.). Dahlem Konferenzen, Berlin.

Manheim, F. T., (1976). Interstitial waters of marine sediments. *In* "Chemical Oceanography." 2nd Ed. Vol. 6 (J. P. Riley and R. Chester, eds). Academic Press, London and New York.

Manheim, F. T. and Sayles, F. L. (1974). Composition and origin of interstitial waters of marine sediments based on deep sea drill cores. *In* "The Sea," Vol. 5 (E. D. Goldberg, ed.). Wiley-Interscience, New York.

Martin, J. M. and Meybeck, M. (1979). *Mar. Chem.*, **7**, 173–206.

Martin, J. M. and Meybeck, M. (1978). Major element content in river dissolved and particulate loads. *In* Biogeochemistry of Estuarine Sediments, Proc. Melreux Symp., Nov. 1976 (E. D. Goldberg, ed.). UNESCO, Paris.

Maxwell, D. T. and Hower, J. (1967). *Am. Mineral.*, **52**, 843–857.

Maynard, J. B. (1981). Chemical mass balance between rivers and oceans: The problem revisited. *In* "Chemical Cycles in the Evolution of the Earth" (R. M. Garrels, B. Gregor, and F. T. Mackenzie, eds). Wiley, New York (in press).

Maynard, J. B. (1976). *Geochim. et Cosmochim. Acta*, **40**, 1523–1532.

McMurtry, G. M. and Yeh, H-W. (1981). *Chem. Geol.*, **32**, 189–205.

Meybeck, M. (1979a). *Rev. Geolog. Dynam. Geograph. Phys.*, **21**, 215–246.

Meybeck, M. (1979b). Pathways of major elements from land to ocean through rivers.

In "Review and Workshop on River Inputs to Ocean Systems." 26–30 March, 1979. FAO, Rome.

Meybeck, M. (1977). Dissolved and suspended matter carried by rivers: Composition, time and space variations and world balance. *In* "Interactions between Sediments and Fresh Water", Junk and Pudoc, Amsterdam.

Meybeck, M. (1976). *Hydrol. Sci. Bull.*, **21**, 265–284.

Mowatt, T. C. and Naid, A. S. (1974). Clay mineralogy and geochemistry of continental shelf sediments of the Beaufort Sea. *In* "The Coast and Shelf of The Beaufort Sea" (J. C. Reed and J. Sater, eds.). The Arctic Inst. N.A., Arlington, Va.

Muffler, L. J. P. and White, D. E. (1969). *G.S.A. Bull.* **80**, 157–182.

Perry, E. and Hower, J. (1970). *Clays Clay Min.*, **18**, 167–177.

Pigott, J. D. (1981). Global Tectonic Control of Phanerozoic Rock Cycling. PhD Thesis, Northwestern University, Illinois (in press).

Ronov, A. (1980). "Sedimentary Cover of the Earth." Nauka Pub. House, Moscow (in Russian).

Russell, K. L. (1970). *Geochim. Cosmochim. Acta*, **34**, 893–907.

Sayles, F. L. (1981). *Geochim. Cosmochim. Acta*, **45**, 1061–1086.

Sayles, F. L. (1979). *Geochim. Cosmochim. Acta*, **43**, 527–545.

Sayles, F. L. and Mangelsdorf, P. C. (1973). *Geochim. Cosmochim. Acta*, **43**, 767–779.

Sayles, F. L. and Mangelsdorf, P. C. (1977). *Geochim. Cosmochim. Acta.*, **41**, 951–960.

Sayles, F. L. and Manheim, F. T. (1975). *Geochim. Cosmochim. Acta.*, **39**, 103–127.

Schink, D. R., Guinasso, N. L., and Fanning, K. A. (1975). *J. Geophys. Res.*, **80**, 3013–3031.

Shaw, D. B. and Weaver, C. E. (1965). *J. Sed. Pet.*, **35**, 213–222.

Sillén, L. G. (1961). The physical chemistry of sea water. *In* "Oceanography (M. Sears, ed.). *Am. Assoc. Adv. Pub.* **67**, 549–581.

Tardy, Y. and Garrels, R. M. (1974). *Geochim. Cosmochim. Acta*, **38**, 1101–1116.

Vail, P. R., Mitchum, Jr., R. M., and Thompson, S. (1977). *Am. Assoc. Pet. Geol. Mem.* **26**, 83–97.

Vinogradov, A. P. (1959). "The Geochemistry of Rare and Dispersed Chemical Elements in Soils." Consult. Bur., Inc., New York.

Weaver, C. E. (1967). *Geochim. Cosmochim. Acta.*, **31**, 2181–2196.

Windom, H. L. (1976). Lithogenous material in marine sediments. *In* "Chemical Oceanography", 2nd Ed. Vol. 5 (J. P. Riley and R. Chester, eds.). Academic Press, London and New York.

Wolery, T. J. (1978). Some Chemical Aspects of Hydrothermal Processes at Mid-Oceanic Ridges. PhD Thesis, Northwestern University, Illinois.

Wolery, T. J. and Sleep, N. H. (1981). Interactions of deep and shallow cycles. *In* "Chemical Cycles in the Evolution of the Earth" (R. M. Garrels, C. B. Gregor, and F. T. Mackenzie, eds.). Wiley, New York (in press).

Wolery, T. J. and Sleep, N. H. (1976). *J. Geol.*, **84**, 249–275.

Wollast, R. (1981). Interactions between major biogeochemical cycles: Marine Cycles. *In* "Some perspectives of the major biogeochemical cycles" (G. E. Likens, ed.). Scope 17. Wiley, Chichester, New York, Brisbane, Toronto.

Wollast, R. (1974). "The silica problem". *In* "The Sea". Vol. 5 (E. D. Golberg, ed.). Wiley-Interscience, New York.

Worsley, T. R. and Davies, T. A. (1979). *Science*, **203**, 455–456.

Yeh, H.-W. and Savin, S. M. (1977). *G.S.A. Bull.*, **88**, 1321–1330.

Ziegler, A. M., Scotese, C. R., McKerrow, W. S., Johnson, M. E., and Bambach, R. K. (1979). *Ann. Rev. Earth Planet. Sci.*, **7**, 473–502.

Natural Water and Atmospheric Chemistry of Silicon

S. R. Aston

Department of Environmental Sciences
University of Lancaster, UK

3.1 INTRODUCTION

Compounds of silicon occur in all natural waters, and may be present either in solution or as suspended solids. By virtue of the abundance of silicon in the crustal rocks (see Chapter 1), silicon compounds are a very major component of the suspended matter derived from weathering processes. In solution, silicon compounds are also fairly abundant, but their concentrations are controlled by processes such as precipitation, biological removal, and recycling.

The concentrations of both particulate and dissolved silicon in natural waters vary widely from one water type to another, and also on a geographical and seasonal basis in response to weathering, productivity, and sedimentation processes. Typical average concentrations of silicon dissolved in river and ocean waters have been reported as 4000 and 1000 µg Si l^{-1} respectively (Riley and Chester, 1971). Silicon distributions in the oceans are discussed in Chapter 4. Miyake (1965) has compiled a summary of the chemical composition of the silica content of rivers and lakes from most parts of the world. The range of SiO_2 content was reported to be 8.6 to 17.9 (wt. %), with a world mean of 11.7 (wt. %).

The variability of dissolved silicon (and other constituents) in river water on a seasonal basis is illustrated by the data on the Moreau River (South

Dakota, USA) compiled by Livingston (1963). Table 3.1 shows the SiO_2 (μgml^{-1}) content of river water from October 1949 to September 1950. The silica content is greatest ($\sim 8\,\mu gml^{-1}$) in Spring and lowest in the winter months ($< 1\,\mu gml^{-1}$). The variations are related to the considerable changes in river discharge, and reflect the input of silicon from the catchment in the high spring movement of water in the system. The seasonal response of silicon concentrations in solution does not, however, occur to the same degree as many other weathering products.

The main purpose of this chapter is to discuss the chemical forms and processes which relate to silicon occurrence in natural waters and the atmosphere. The importance of silicon in its role as a nutrient in natural waters is discussed in Chapter 4, the solubility of solid phases in natural waters is mentioned in several chapters, and its behaviour in the sediments underlying the waters is dealt with in Chapter 5. Consequently, the present discussion will be restricted in the most part to abiological aspects and will only include discussion of sedimentary and particulate materials when they are in suspension in water bodies or in the atmosphere.

In the following sections, the silicon geochemistry of natural waters has been divided up to deal with the major types of natural waters, i.e. fresh water, ocean water, and brackish water. Before each of these water types is discussed, Section 3.2 will deal with the general problem of the chemical

TABLE 3.1   The temporal changes in dissolved silicon (as SiO_2) in the Moreau River, South Dakota (data from Livingston, 1963).

Date of sample	SiO_2 (μgml^{-1})	Date of sample	SiO_2 (μgml^{-1})
1949		*1950*	
Oct. 1–3	0.2	Apr. 18–20	3.1
Oct. 4–12	0.3	Apr. 21	2.1
Oct. 13–26	1.3	Apr. 22–26	2.1
Oct. 27–31	1.0	Apr. 27	1.5
Nov. 1–30	0.8	Apr. 28–May 15	2.0
Dec. 1–4	0.4	May 16–31	1.5
		June 1–30	0.5
1950		July 1–31	0.5
Mar. 6–9	3.0	Aug. 1–5	0.6
Mar. 10	1.5	Aug. 6–8	2.1
Mar. 14–Apr. 1	3.2	Aug. 9–31	0.7
Apr. 3	8.1	Sept. 1–19	0.6
Apr. 4–6	6.4	Sept. 20–24	1.9
Apr. 7	5.4	Sept. 25–30	0.7
Apr. 11–14	5.0		
Apr. 15–17	5.6		

speciation of silicon compounds in the natural water environment. In conclusion, Section 3.6 is concerned with the presence of silicon in the Earth's atmosphere, and its supply and removal in the atmospheric environment.

3.2 CHEMICAL SPECIATION OF SILICON IN NATURAL WATERS

All attempts to understand geochemical and biogeochemical processes depend upon a knowledge of the chemical forms, or species, which are present in the natural environment. Despite this well-recognised requirement, it is true to say that for many elements their speciation in the natural environment is still a subject of comparative ignorance and debate. In recent years, an increasing amount of attention has turned to speciation problems in natural waters. However, a good deal of this research has been and continues to be concerned with trace elements, particularly transition metals and Group IIb elements. These elements are of particular importance as pollutants in natural waters. Rather less attention has been focused on the more abundant elements, including silicon and other nutrients. To some extent this state of affairs can be understood because, in addition to the applied problem of water pollution by certain trace elements, these are elements of interest to the chemist as a result of their ready ability to form organic and inorganic complexes, to undergo adsorption/desorption reactions on reactive surfaces, e.g. sediment grains, and to take part in various redox reactions.

When discussing silicon in natural waters most of the literature commonly refers to the form of silicon present in solution as "silicate". This anion, SiO_4^{4-}, being derived from weathering reactions of silicate and aluminosilicate minerals. A simplified representation of the weathering of the common mineral potassium feldspar is:

$$4KAlSi_3O_8 + 22H_2O \rightarrow 4K^+ + 4OH^- + Al_4Si_4O_{10}(OH)_8 + 8H_4SiO_4$$

K-feldspar

This reaction is typical of many weathering reactions in which the products of weathering are a solid clay mineral, e.g. kaolinite, and silicic acid in solution. Undissociated silicic acid is the stable form of the silicate species at pH values less than 9 (Bruevich, 1953; Sillen, 1961; and Siever, 1971). When the pH rises to beyond 9 some dissociation is possible and should occur.

$$H_4SiO_4 \rightleftharpoons H_3SiO_4^- + H^+$$

$$H_3SiO_4^- \rightleftharpoons H_2SiO_4^{2-} + H^+$$

Dissociation of silicic acid is, of course, unlikely in most natural waters because their pH is maintained at values below 9. Sea water usually has a pH of 8–8.1, while fresh waters have a variety of pH values typically in the range of 4.5–7. The dissociation reactions shown above are only plausible when alkaline lakes are considered. The dissociation reactions of silicic acid have been studied by Lagerstrom (1959) for $NaClO_4$ solutions. The first and second dissociation constants of silicic acid appear to be about 10^{-10} and 10^{-12} respectively.

Since the pH of virtually all natural waters is 9, the presence of undissociated silicic acid is of prime importance to silicon speciation. This relatively simple situation is, however, complicated by the question of whether or not the silicic acid is present in a monomeric or polymeric form. The pure and applied colloid chemistry of silicon has been reviewed by Iler (1955), and from the standpoint of natural waters it is interesting to note that the silicic acid may, at least in theory, exist in both monomeric and polymeric species. Many of the determinations of silicic acid in natural waters have been made by the classical silico-molybdate method. The yellow complex formed in this technique is the result of the reaction of monomeric and dimeric silicic acid alone. As a result, the results reported are specific to these species while any polymers present are neglected. Since 1955, the method of Mullin and Riley (1955) has also been extensively used for natural water analysis. This technique employs the reagent p-methylamino phenol sulphate. When this reagent is present as a strong reducing agent, an intensely blue silicomolybdate complex is formed. This procedure can be modified to allow the determination of all polymeric forms of silicic acid rather than the monomer and dimer only. The determination of all forms of silicic acid is achieved by a pretreatment of the natural water sample by heating with sodium hydroxide in a polytetrafluoro-ethane vessel (Morrison and Wilson, 1967). The application of the above analytical methods should allow the extent of polymerization in natural waters to be estimated. However, the question of whether natural polymeric forms of silicic acid exist is not clearly resolved.

Kruaskopf (1956) has studied the dissolution and precipitation of aqueous silicic acid at low temperatures. He found that polymerization was a very slow process at ordinary, i.e. earth surface, temperatures. Krauskopf (1956) provided evidence (Table 3.2) that the coagulation of a silica sol required days or weeks at near neutral pH values, and under acid conditions the process was greatly retarded. At higher temperatures, which are of little interest except in the case of hot springs, the polymerization appeared to be much faster. Krauskopf (1956) concluded that the formation of a silica sol was possible by evaporation, co-precipitation with other colloids, and in the presence of fairly concentrated solutions of electrolytes. Precipitation of silica by electrolytes was reported as being fastest in basic solutions and extremely slow at pH values <6. Furthermore, Krauskopf concluded that for undersaturated true solutions of silicic acid, the formation of silica gels was not influenced by the presence of electrolytes or other colloids. When the conditions of natural waters are considered, it is evident from the above observations that silicic acid should not readily form silica sols by polymerization reactions.

The extensive collection of seawater samples taken from various oceanic areas by Lisitsyn *et al.* (1966) has been used to examine the occurrence of amorphous and colloidal silica in ocean waters. These authors found that, except for very small amounts, colloidal polymerized silicic acid was absent. Even the very small amounts of colloidal material found were shown to be of biogenic rather than chemical origin.

Burton *et al.* (1970) carried out an investigation into the presence of the so-called "reactive" and "non-reactive" fractions of dissolved silicon in fresh and marine waters. The terms reactive and non-reactive refer, in essence, to the monomeric silicon and polymeric silicon species in solution in sea water. The meaning of the term reactive as applied in this sense is that this form will participate in the biological cycling of silicon in sea water (and presumably other natural waters). In contrast, it is also assumed that polymeric forms are

TABLE 3.2 The formation of a silica gel by the polymerization of dissolved silicon at various conditions (data from Krauskopf, 1956).

	I	II	III	IV	V
pH during experiment	7.7–8.3	7.3–7.9	4.8–5.3	6.9–7.3	8.0–9.1
Original SiO_2 (total)	320	975	171	171	171
Original SiO_2 (dissolved)	284	544	162	162	164
Dissolved SiO_2 after:					
1 day		172	163	163	164
2		167			
3	173				
8			164	163	165
10	148				
12		135			
19			165	163	161
24	131				
26		130			
38	122				
39		130			
52	115				
68	113				
128			160	156	149

Concentrations were ppm SiO_2, temperature range 22–27°C.

not available for biological uptake and utilization. The terms reactive and non-reactive have remained in common use by biological oceanographers.

In their experiments, Burton *et al.* (1970) used samples of deep North Atlantic sea water and river waters from four locations in southern England. These authors found no evidence (Table 3.3) for any presence of non-reactive silicon in any of their water samples. This conclusion was based on the analysis of the waters by the method of Mullin and Riley (1955) for reactive silicon, and its modification to determine total silicon by the use of Morrison and Wilson's (1969) technique. In this manner, non-reactive silicon was defined as the difference between the silicon concentrations as found by the two methods.

TABLE 3.3 Typical results for reactive and total dissolved silicon from replicate analyses of samples from three rivers in Hampshire, England (data from Burton *et al.*, 1970).

River	Reactive dissolved silicon (μgl^{-1})	Total dissolved silicon (μgl^{-1})
Test	4728	4734
Itchen	4827	4824
Beaulieu	5533	5532

In addition to the above analyses, Burton *et al.* (1970) investigated the possible depolymerization of the so-called unreactive silicon when added to natural waters. This piece of work is of great interest to the whole question of whether or not silicon can exist in a polymeric form under natural water conditions. The results obtained by Burton *et al.* (1970) showed that depolymerization of polymerized silicic acid occurred in both saline and fresh waters. The rate of depolymerization was faster in sea water than in fresh waters. For example, $5 \, mgl^{-1}$ of polymeric silicon was completely de-polymerized in sea water after 18 hours. In distilled water the same concentration of polymeric silicon had a residual 3% of polymeric silicon remaining after 12 days. An important point made by the authors was that the rapid depolymerization of silicon in sea water followed zeroth order kinetics, and was probably a surface-limited reaction. They concluded that this precluded the direct extrapolation of their reaction rate data to the natural environment.

In summary, it may be stated that the dissolved form of silicon present in most natural waters is undissociated silicic acid (H_4SiO_4). In certain highly alkaline waters (e.g. some evaporite lakes) with pHs >9 there may be significant dissociation. The question of whether or not monomeric silicic acid is converted to polymeric forms under environmental conditions is still somewhat obscure, but there is some evidence to suggest that polymeric forms are not stable in natural waters. This topic remains one with much potential for further research.

3.3 SILICON IN FRESH WATERS

The pH of fresh waters is such that dissociation of silicic acid, which only becomes appreciable at pH > 9, should be absent and all such waters should then contain monomeric undissociated H_4SiO_4 as the dissolved silicon species (see Section 3.2). There is little evidence to suggest that any polymers of silicic acid will occur in fresh waters, e.g. lakes and rivers (Hem, 1970; Burton *et al.*, 1970).

The silicon content of rivers and lakes is probably still best exemplified by the compilation of data made by Livingston (1963), who brought together an enormous body of information on the chemical composition of fresh waters from all over the World. Table 3.4 summarizes the data of Livingston (1963) on the occurrence of dissolved silicon in fresh waters from six major continental areas and the overall global data based on these compilations. The range of the average SiO_2 content for the global data is rather small (3.9 to 23.2 ppm), especially when compared with the ranges of concentrations found for other major and trace constituents of fresh waters on a worldwide basis.

The relative lack of variation of dissolved silicon concentrations in an individual river system has already been noted (see Section 3.1), and this is the picture which has been observed for several rivers. For example, the work of Gibbs (1972) showed that in the world's greatest river system, the Amazon,

TABLE 3.4   Summary of data on the silicon content of fresh waters on a global basis (data from Livingston, 1963).

Area	SiO_2 content (ppm)
North America	9.0
South America	11.9
Europe	7.5
Asia	11.7
Africa	23.2
Australia	3.9
World	13.1

the range of concentration of dissolved SiO_2 was only 9 to 12 mgl^{-1} even though there are substantial changes in the discharge pattern. These observations have led Edwards and Liss (1973) to conclude that a buffering effect on the dissolved silicon in fresh waters may be operating. These authors pointed out that while the dissolution of solid phases and biological activity are important in the geochemical behaviour in silicon in natural waters, these processes cannot be used to explain fully the low temporal and spatial variability found for dissolved silicon concentrations in fresh waters. Edwards and Liss (1973) suggested that sorption and desorption reactions play an important role in the buffering of dissolved silicon concentrations in rivers and lakes. They cited the observations and evidence for a sorption reaction control of the silicon content of soils, where oxides of iron and aluminium appear to act as adsorption/desorption sites for silicon (Jones, 1963; Beckwith and Reeve, 1964). The role of the sorption/desorption of silicon in the marine water–sediment system has been investigated in some detail (see Chapters 4, 5, and 6). For fresh waters, especially lakes, data is also available on the exchange of silicon between bottom deposits and the water body (see, e.g. Gardiner, 1941; Mortimer, 1941; Kato, 1969; Burns and Ross, 1972; Møller-Anderson, 1974; Dickson, 1975; Youngman et al., 1976; and Rippey, 1980). The latter work of Rippey on the freshwater lake Lough Neagh, Northern Ireland, provides a very detailed account of the exchange of silicon between the solid-soluble phases. Rippey (1980) concluded that the exchange reaction was very dependent on temperature, and that the kinetics of the process of silicon release (mainly from diatom remains) are best explained by a first-order dissolution reaction and a parabolic readsorption by kaolinite and/or illite in the sediments. While this, and other, quantitative measurements of the influence of solid phases on the control (and buffering) of silicon in fresh waters are useful, further research into the mechanism(s) of the buffering process(es) seems probably the most fruitful line to follow in our understanding of non-biological controls on the aquatic geochemistry of silicon.

The role of silicon in the biology of fresh waters has long been recognized, with diatoms capable of using dissolved silicon at low concentrations and providing the seasonal cycle of silicon in many freshwater bodies. Early examples of the biological use and seasonal cycles of freshwater bodies are available (see, e.g. Yoshimura, 1930; King and Davidson, 1933; and Mortimer, 1941). The biological cycles of silicon in lakes have been reviewed by Macan (1970), Hutchinson (1975), and Cole (1979). Several aspects of the biological cycle of silicon are discussed with reference to freshwater lakes in Chapter 5, and are not discussed further here.

3.4 SILICON IN ESTUARINE AND BRACKISH WATERS

When sea water and river water are mixed together in estuaries and other coastal embayments chemical reactions may be expected to occur. The behaviour of silicon during these mixing processes has received a considerable amount of attention, but the picture which emerges is far from clear. Since other chapters deal with various aspects of silicon in other natural waters, here the estuarine chemistry of silicon is developed in more detail than fresh and sea water. Before considering the behaviour of silicon in estuaries, it is important to summarize the framework within which the various studies have been carried out.

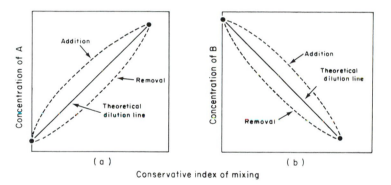

FIG. 3.1   An idealized representation of the relationship between the concentration of a dissolved component and a conservative index of mixing, e.g. chlorinity (from Liss, 1976); (a) for a component A whose concentration is greater in sea water than in river water, (b) for a component B whose concentration is greater in river water than in sea water.

Essentially, two methods of study have been used. First, the collection of samples to represent river water through intermediate salinity waters to sea water. This range of samples from an estuary represents the silicon content of the two *end members*, river and sea waters, and a selection of natural mixtures. Using these samples and an index of conservative mixing, the loss or addition of dissolved silicon during estuarine processes can be evaluated. Typically, the index of conservative mixing would be salinity or chlorinity. The idealized representation of mixing curves is shown in Fig. 3.1. When conservative behaviour occurs, the data points should lie on, or very near to, the straight line joining the end members.

Although this first approach is simple in concept, there are important problems associated with its use in real estuarine situations. Foremost among these difficulties is the definition of the end-member concentrations which are of the utmost importance in the construction of the mixing curve. Problems which occur include:

(i) the variability of end-member concentrations;
(ii) the possibility of multiple source inputs;
(iii) the analytical definition of "dissolved" concentrations; and
(iv) the lack of mechanistic interpretation which can be applied to dilution curve data.

The second method of study of estuarine mixing processes is the laboratory simulation approach, where river and sea water are mixed together experimentally. This type of approach also has some important disadvantages, such as the great difficulties involved in any laboratory experiment which tries to model a real environment. The limitations of these experiments is discussed in more detail below.

Table 3.5 shows the list of field and laboratory studies aimed specifically at the conservative–nonconservative behaviour of silicon in estuarine and brackish waters. The earliest studies were those of Maéda and his co-workers who investigated the behaviour of dissolved silicon in some Japanese estuaries (Maéda, 1952, 1953; Makimoto et al., 1955; Maéda and Tsukamoto, 1959; and Maéda and Takesue, 1961). These workers found that a linear relationship existed between chlorinity and dissolved silicon concentrations. Their data are of limited use because of the very limited range of chlorinity examined. Another early study was that of Bien et al. (1958) on silicon in the Mississippi estuary. In contrast to the Japanese results, these authors reported large-scale silicon removal. Figure 3.2 contrasts the typical results reported by Bien et al. (1958). The Mississippi data have been reinterpreted by Schink (1967) who suggested that there is only evidence of $\leqslant 10\%$ silicon removal from solution in the Mississippi. In total, field studies on dissolved silicon behaviour in estuaries have been reported for locations from the USA, China, Japan, India, UK, and Venezuela. It is apparant from the findings summarized in Table 3.5 that there is considerable disagreement between these respective studies. The reasons for these discrepancies may be various. Different estuaries with their widely varying conditions of catchment and freshwater input, mixing lengths and times, and silicon concentrations in the end-member solutions, might be expected to show different pictures of silicon behaviour. On the other hand, the differences may be attributed to different techniques adopted by the investigators, poor definition of end members and/or choice of conservative index, and lack of appreciation of complicating factors, e.g. biological utilization of silicon (see below). Reviews of silicon behaviour in estuaries as observed in the field have been compiled by Liss (1976) and Aston (1978). The overall conclusion which

TABLE 3.5 A compilation of field and laboratory studies aimed specifically at the conservative–nonconservative behaviour of silicon in estuarine and brackish waters.

Estuary	Finding	Reference	River concentration of SiO_2 (mg l^{-1})
Various, Japan	Linear relationship between Si and chlorinity; numerous surveys generally over a limited chlorinity range	Maéda (1952, 1953), Makimoto et al. (1955), Maéda and Tsukamoto (1959), Maéda and Takesue (1961)	—
Mississippi, USA	10–20% removal	Data of Bien et al., (1958) interpreted by Schink (1967)	6
Columbia, USA	No significant removal	Stefánsson and Richards (1963), Park et al. (1970)	12
Jiu-Long, China	Considerable removal (20–30%)	Fa-Si et al. (1964)	16
Kiso and Nagara, Japan	Linear relationship between Si and chlorinity; surveys in both estuaries over a limited chlorinity range	Kobayashi (1967)	—
Southampton Water, UK	No significant removal	Burton et al. (1970b)	10
Hamble, UK	10% removal	Burton et al. (1970b)	12
Vellar, India	10% minimum removal (see text)	Burton (1970)	39
Chikugogawa, Japan	Considerable removal (20–30%[a])	Hosokawa et al. (1970)	30
Conway, UK	10–20% removal	Liss and Spencer (1970)	3
St. Marks, USA	Considerable removal	Stephens and Oppenheimer (1972)	12
Alde, UK	25–30% removal	Liss and Pointon (1973)	7
Orinoco, Venezuela	Inconclusive, lack of data on fresh water end member	Fanning and Pilson (1973)	—
Savannah, USA	Probably no significant removal	Fanning and Pilson (1973)	9
Mississippi, USA	7% removal	Fanning and Pilson (1973)	6
Merrimack, USA	No significant removal	Boyle et al. (1974)	5
Various, Scotland	No significant removal	Sholkovitz (1976)	0.3–1.3

[a] Extent of removal not estimated quantitatively. Figure given is from interpretation by Burton and Liss (1973) (after Liss, 1976).

can be drawn on this topic is that little is really known, and that this is an aspect of silicon geochemistry which is worthy of further research.

The laboratory data on silicon behaviour in brackish waters is not much help in resolving the problems discussed above. Krauskopf's (1956) suggestion that silicic acid may be converted to a polymeric colloidal form during the mixing of river water with sea water was further investigated by Burton *et al.* (1970) and Seiver (1971). These workers found no evidence for the

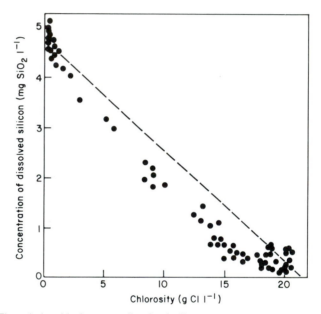

FIG. 3.2 The relationship between dissolved silicon concentration and chlorosity in the Mississippi Delta (June 1953). The broken line represents the theoretical dilution curve (after Bien *et al.*, 1958).

presence of a polymeric form. Furthermore, as stated above, Burton *et al.* (1970) showed that when polymeric silicic acid is placed in sea water it rapidly depolymerizes to monomers of silicic acid. This suggests that if silicon removal does occur in some or all estuaries the mechanism is not polymerization and the formation of colloids, but some other unknown process.

The most detailed laboratory study on estuarine mixing and conservative/non-conservative behaviour of silicon and other elements is that of Sholkovitz (1976). In this study the *products* of river and sea water mixing in the laboratory were studied. This product approach was adopted because

Sholkovitz (1976) criticized the studies based on the deviations from dissolved concentrations on four counts:

(1) end member definition;
(2) multiple source inputs;
(3) defining of particulate matter; and
(4) lack of mechanistic approach.

The product approach itself has several limitations, and these have been discussed by Aston (1980). These problems may be very briefly summarized as:

(i) in common with all laboratory simulations there must be doubts as to the representative nature of kinetic and thermodynamic conditions;
(ii) the ability to define and collect particulate products;
(iii) difficulties simulating the influence of pre-existing particles on the reactions; and
(iv) the influence of vessel surfaces, e.g. glass reaction flasks, which may play a significant role.

Setting these difficulties aside, the results obtained by Sholkovitz (1976) indicate that silicon removal from solution during estuarine mixing should be a mere 3–6% of the total dissolved fraction (see Fig. 3.3).

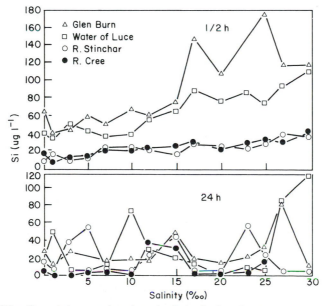

FIG. 3.3   Silica flocculation as a function of salinity for four Scottish river systems. Results for $\frac{1}{2}$-h and 24-h experimental runs are shown (after Sholkovitz, 1976).

Earlier, and more simple, experiments in the laboratory have been used to examine silicon behaviour during estuarine mixing (Bien *et al.*, 1958; Liss and Spencer, 1970; and Fanning and Pilson, 1973). Bien *et al.* (1958) demonstrated silicon removal from solution during the experimental mixing of river and sea waters. Their results showed that such removal was only significant when suspended particles were present. This is pertinent to the results of Sholkovitz (1976) where prefiltered waters were used to remove the existing suspended materials. Bien *et al.* (1958) demonstrated that the presence of natural suspended solids, alumina, and bentonite particles, were all effective in rapidly removing dissolved silicon in the mixing experiments. Silicon removal was a function of the concentration of suspended solids, and the most effective removal occured when natural suspended matter was employed. Bien *et al.* concluded that the suspended materials provided active surfaces for adsorption of silicate. Similar results were obtained by Liss and Spencer (1970), and these workers showed that silicate desorption from partialates does not occur under their laboratory mixing experiment conditions. The third group of experiments on river–seawater mixing performed by Fanning and Pilson (1973) gave results which do not agree with the findings of Bien *et al.* (1958) and Liss and Spencer (1970). These later studies showed no significant losses of silicon from solution in experiments using Mississippi River water and seawater samples. Fanning and Pilson (1973) concluded that the differences between their results and those from the earlier experiments may be due to differences in the experimental conditions, in particular temperature and analytical methods for the deformination of dissolved silicon. Fanning and Pilson (1973) have pointed out some experimental difficulties in the analytical techniques used by Bien *et al.* (1958), which suggest that this older data may be unreliable. With regard to the discrepancies between their results and those of Liss and Spencer (1970), Fanning and Pilson (1973) have drawn attention to the fact that their experiments were conducted at 17–28°C while those of Liss and Spencer (1970) were performed at 10°C. They proposed that the lowering of the experimental temperature by 7–18°C may be sufficient to reduce the "solubility" of various silicate minerals and thus cause some precipitation. No concrete evidence was available on this aspect, but it may be invoked as a possible cause of the conflicting results produced by the various experiments.

In conclusion, the whole question of whether or not the silicon behaves in a conservative or non-conservative manner in estuaries and brackish water remains fairly open. For this reason, this aspect of the aquatic geochemistry of silicon is one with potential for further research. The following points are intended as a guide to those problems which remain to be clearly resolved, and some comments on possible useful approaches:

(1) Various workers have used different estuaries and analytical methods in the field approach to this problem. A wide-ranging study on several estuaries of different physical and geological characteristics undertaken by a single research group using a unified approach to sampling and analysis may help to eliminate many uncertainties.

(2) A similarly unified approach to laboratory studies of river–seawater interactions, with and without suspended solids, could be very illuminating. This study should involve loss from solution and *product* approaches, and be extended to a wide range of temperatures.

(3) Detailed seasonal field studies on silicon behaviour in an estuary should help to distinguish any possible abiological and biological removal processes, and illustrate their relative importance (see below).

(4) Although the results of Burton *et al.* (1970) on the occurrence and behaviour of polymeric silicic acid are interesting, there appears to be scope for further research into the role of dimerization and polymerization of silicic acid in natural waters of various types.

(5) The selection of indices of conservative mixing for estuarine studies requires critical appraisal, and novel methods of interpreting the mixing of river and sea water should be tried out. For example, Boyle *et al.* (1974) have suggested that the $^{18}O/^{16}O$ isotopic ratio may eliminate many of the problems encountered with the application of conventional indices.

Although it has been stated earlier that this chapter will tend towards abiological aspects of silicon in natural waters, it is impossible (as stated for fresh water in Section 3.2) to avoid biological influences. The biological removal and cycling of silicon in estuarine waters have recently been reviewed (Aston, 1980) and some of the more important conclusions are given here. Silicon will be utilized by siliceous diatoms, silico–flagellates, and sponges in brackish waters, and this removal will depend on several factors in the estuarine environment, e.g. populations present, rate of growth of the organisms, physical influences on the organisms, and the resolution of silica from dead organisms. The biological influence on dissolved silicon in estuarine waters is, of course, a seasonal one (except in tropical regimes), with more rapid and extensive use of silicon in spring and summer. Little field data on biological influences on silicon distribution in estuaries exist, but some detailed surveys do exist, see e.g. Peterson *et al.* (1975). The regeneration of dissolved silicon species from the decomposition of dead siliceous organisms in brackish waters is not well characterized, and most information is based on rather old laboratory studies (Atkins, 1945; Harvey, 1955; and Jørgensen, 1955). A high degree of variability of resolution of silica frustules has been reported, and Aston (1980) has concluded that in estuaries the resolution process is not likely to be very quantitatively important.

Figure 3.4 represents the silicon cycle for a hypothetical estuary based upon abiological and biological transfer processes. The quantitative data needed to transform this into a useful model of silicon in the estuarine environment does not as yet exist for the most part. Only in the case of the conservative/non-conservative behaviour of silicon in estuaries has there

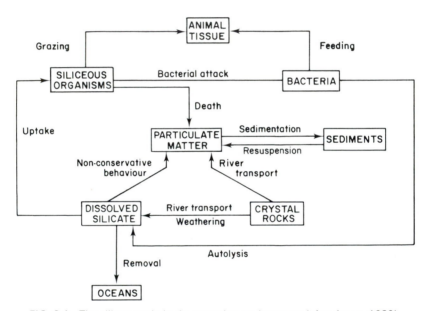

FIG. 3.4   The silicon cycle in the estuarine environment (after Aston, 1980).

been a reasonable attempt to model the field data in a numerical approach (Rattray and Officer, 1979). Using the data on the distribution of dissolved silica in the San Francisco estuarine system (Peterson *et al.*, 1978) and a hydrodynamic model for two-dimensional, gravitational circulation (Festa and Hansen, 1976) these authors have tested the relation between the distribution of silicon in solution and the distribution of salinity analytically from conservation equations. The relation developed for the dissolved silicon distribution from the silicon and salt conservation equations was compared, with satisfactory agreement, against a numerical model determination of the same problem. This approach is a useful step towards a more rigorous and systematic approach to the estuarine behaviour of dissolved silicon and other chemical parameters.

3.5   SILICON IN MARINE WATERS

The geochemistry and biological utilization of silicon in seawater are discussed in detail in other parts of this book (Chapters 4, 5, and 6). For this reason very little attention is given here to the marine cycle of silicon, and it is merely re-emphasized that in this natural water the predominant species of silicon found in solution will be undissociated silicic acid. As with other natural waters, there is no real evidence to lead to the conclusion that the dissolved silicon should be in polymeric forms (see Section 3.2), and the supply and removal of silicon in the oceans is controlled by those processes discussed in other chapters, and the important role of the oceans in the global cycle of the element is dealt with in Chapter 2).

3.6   SILICON IN THE ATMOSPHERE

Silicon is present in the Earth's atmosphere as a consequence of several processes which may be summarized as follows:

(i) the suspension of silicon containing particles of inorganic and biological origin from the Earth's surface;
(ii) the release of silicon compounds from anthropogenic sources, e.g. industry and fuel consumption to the atmosphere; and
(iii) the introduction of silicon radionuclide(s) from cosmic ray interactions in the upper atmosphere.

Each of these sources is discussed below, and in the first situation, i.e. suspension of particles, the processes controlling the introduction distribution and fate of the material is considered. Little information is available on the latter two sources compared to the former.

Particulate material derived from the earth's surface, especially soils, is a potential source of silicon in the atmosphere both as particulate material and in rain-water. The resuspension of surface particles into the atmosphere is a complex physical process and will not be considered here. Suffice to say that it is dependent on the source materials available, the arid nature of individual regions, and prevailing wind conditions in the boundary layer. Clearly, each of these factors is subject to considerable variation over the surface of the earth. The removal of surface particles from soil and their resuspension into the lower troposphere is not expected to result in any significant chemical changes in the silica and silicate minerals present. Hence, atmospheric dust is of interest in the geochemistry of silicon from the point of view of a transport path in the surface environment rather than an important stage in the geochemistry or biogeochemistry of the element.

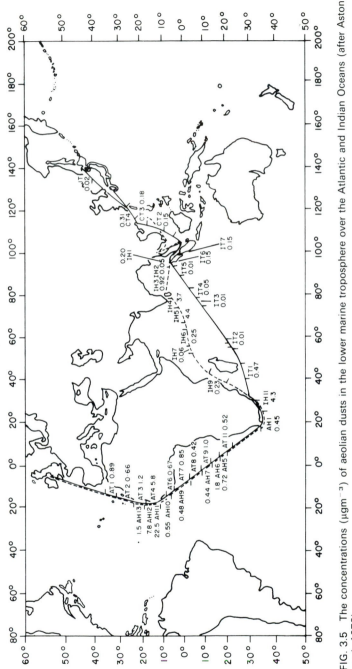

FIG. 3.5 The concentrations ($\mu g\,m^{-3}$) of aeolian dusts in the lower marine troposphere over the Atlantic and Indian Oceans (after Aston et al., 1973)

The transport of atmospheric dusts, and thus silicon containing minerals, was probably first noted by the famous naturalist Darwin (1846). During the voyage of the *Beagle* he recorded the presence of reddish-brown dust in the North Atlantic far from land masses, and recognized that such dusts may constitute a significant pathway for sedimentary materials between the continents and the oceans. The detailed investigation of atmospherically (aeolion) transported commenced with the studies of Radczewski (1939) who characterized rounded quartz grains coated with red-brown iron oxide in the marine sediments off the coast of West Africa. These grains are atmospherically transported from the adjacent Sahara Desert in the prevailing Trade Winds. Rex and Goldberg (1958) extended the studies on marine sediments and were able to show the worldwide importance of atmospherically transported silica grains in the accumulation of oceanic deposits. However, it is only within recent years that the quantitative assessment of this transport process has been investigated in any real detail. The geographical distribution of atmospheric dusts, dominated by quartz, has been mainly investigated in oceanic regions in relation to continental desert source areas and wind systems. The dust loadings in the lower troposphere vary considerably and this is exemplified by the data of Aston *et al.* (1973) in Fig. 3.5. The range of dust loadings was reported as 0.21 to 7.7 μg m^{-3}, and the spatial variations have been related to source areas in this and other studies. A review of the contribution of aeolion material (and consequently quartz which is one of its main constituents) to sediments has been provided by Windom (1975). Table 3.6 shows some data on the average quartz distribution in airborne dusts and associated marine sediments in the world ocean. This author noted the difficulties which arise in trying to estimate the quantitative contribution of atmospherically transported materials to the sedimentary cycle. However, it can be concluded that the process is more significant in deep ocean areas than in shallow-water marine areas where river transport of silica and other sedimentary solids is more dominant. Judson (1968) estimated the aeolian contribution to the oceans to be 0.1 to 0.4 $\times$ 10^9 metric tons per year, which is in good agreement with the calculations of Goldberg (1971) who found that atmospheric transport of silicate materials from the land to the oceans should be about 0.1 $\times$ 10^9 metric tons per year. Windom (1975) has pointed out that the aeolian contribution to sediments in the deep ocean is probably in the range of 10 to 30% of the total deposition, while for all ocean areas the quantitative importance of this pathway is reduced to only a few per cent of the total flux to the sea-bed. It may be concluded that in the deep ocean the atmospheric transport path from the continents is a significant pathway for silicon in the geochemical cycle. See also Chapter 3, Section 2.2.2.

The mechanism of removal of the quartz grains from the atmosphere has been studied by Delany *et al.* (1967) who concluded that in oceanic regions the fall-out of particles from the troposphere is largely a result of gravitational settling and scavenging during precipitation of rain. The residence time of silica dusts in the atmosphere is not long, probably in the region of 10 days.

There appears to be relatively little published information on the concentration of silicon in rain-waters. Turekian (1969) cites some unpublished data from Sugawara (1967) in which the mean silicate concentration in rain-waters is given as $0.00083 \, \mu g \, Si \, l^{-1}$. This is extremely small compared to the silicate content of fresh waters at the Earth's surface, and it may be concluded that silicon is a quantitatively very important constituent

TABLE 3.6 Average quartz distribution in dusts from marine atmospheres and associated marine sediments (data from Windom, 1975).

Region	Quartz content (wt. %)
North Atlantic	
(1) Dusts	
N. African Coast	9–15
Barbados	10
(2) Sediments	
East of Mid-Atlantic Ridge	14
Ridge and Flanks	10
West of Ridge	8
North Pacific	
(1) Dusts	
Eastern N. Pacific	11
(2) Sediments	
North Pacific	10
South Pacific	
(1) Dusts	no data
(2) Sediments	
Western S. Pacific	9
East Pacific Rise	20
Indian Ocean	
(1) Dusts	
Northern	9
South-western	10
Bay of Bengal	11
(2) Sediments	
Total average	17
Bay of Bengal	13

of rain-waters. Also, the role of rain-water in the geochemical transport and pathways of silicon is probably insignificant.

Little information is also available on the occurrence of organic volatile silicon compounds in the atmosphere. Graedel (1978) has summarized the occurrence of organic forms of various metals and metalloids in the lower atmosphere, and has included some data on silicon compounds reported by Pellizzari *et al.* (1976). Table 3.7 lists the silanes which have been found at trace concentrations in air. These substances, which are presumed to be derived from industrial processes, are unstable in the atmosphere and are readily removed by oxidation reactions.

In the ionosphere the presence of silicon ions has been reported (McEwan and Phillips, 1975) and may be derived from meteorite ablation. In a narrow layer at about 100 km above the earth the presence of Si^+ ions has been reported as a common feature of the daytime ionosphere (Narcisi, in press).

Silicon-32, a radioactive isotope of half-life 7.1×10^2 y, occurs in the atmosphere as a consequence of the interaction of cosmic rays with argon. The factors controlling the production of ^{32}Si are complicated, and involve various physical aspects of the energy spectrum of the incoming cosmic

Table 3.7 Silane compounds identified at trace concentrations in air (data from Pellizzari *et al.*, 1976).

Methylsilane	$CH_3Si—CH_3$
Tetramethylsilane	$(CH_3)_4—Si$
Hexamethylcyclotrisiloxane	
Tetradecamethylhexasiloxane	$(CH_3)—Si—O—[Si(CH_3)_2—O]_4—Si(CH_3)_2$
Octamethylcyclotetrasiloxane	

radiation flux (Lal *et al.*, 1958). The rate of production is higher at the geomagnetic poles than at the equator. The estimated steady state activity of ^{32}Si in the troposphere is 8×10^{-9} disintegrations $cm^{-2} s^{-1}$. The contribution of other potential sources of ^{32}Si in the atmosphere, e.g. nuclear weapon tests, appears to be negligible (Kharkar *et al.*, 1966). Most data on the production and occurrence of ^{32}Si in the atmosphere have been inferred from studies on this radionuclide in the oceans, marine organisms and sediments. This topic has been reviewed by Burton (1975).

REFERENCES

Aston, S. R. (1978). *In* "Chemical Oceanography", Vol. 7 (J. P. Riley and R. Chester, eds.). Academic Press, London and New York.
Aston, S. R. (1980). *In* "Chemistry and Biogeochemistry of Estuaries" (E. Olausson and I. Cato, eds.). John Wiley, London.
Aston, S. R., Chester, R., Johnson, L. R., and Padgham, R. C. (1973). *Mar. Geol.*, **14**, 15.
Atkins, W. R. G. (1945). *Nature, Lond.*, **156**, 446.
Beckwith, R. S. and Reeve, R. (1964). *Australian J. Soil Res.*, **1**, 157.
Bien, G. S., Contois, D. E., and Thomas, W. H. (1958). *Geochim. Cosmochim. Acta*, **14**, 35.
Boyle, E., Collier, R., Dengler, A. T., Edmond, J. M., Ng, A. C., and Stallard, R. F. (1974). *Geochim. Cosmochim. Acta*, **38**, 1719.
Breuvich, S. V. (1953). *Iz v. Akad. Nauk. S.S.S.R. Ser. Geol.*, **4**, 67.
Burns, N. M. and Ross, C. (1972). Canada Centre for Inland Waters, Paper No. 6. Project Hypo.
Burton, J. D. (1975). *In* "Chemical Oceanography", Vol. 3 (J. P. Riley and R. Chester, eds.). Academic Press, London and New York.
Burton, J. D., Leatherhead, T. M., and Liss, P. S. (1970). *Limnol. Oceanogr.*, **15**, 473.
Cole, G. A. (1979). "Textbook of Limnology". 2nd Edn. C. V. Mosby, St. Louis.
Darwin, C. (1846). *Q. J. Geol. Soc. Lond.* **2**, 26.
Delany, A. C., Delany, Audrey C., Parkin, D. W., Griffin, J. J., Goldberg, E. D., and Reinmann, B. E. F. (1967). *Geochim. Cosmochim. Acta*, **31**, 885.
Dickson, E. L. (1975). *Freshwater Biol.* **5**, 1.
Edwards, A. M. C. and Liss, P. S. (1973). *Nature, Lond.*, **243**, 341.
Fanning, K. A. and Pilson, M. E. Q. (1973). *Geochim. Cosmochim. Acta*, **37**, 2405.
Festa, J. F. and Hansen, D. V. (1976). *Estuar. Coastal Mar. Sci.*, **4**, 309.
Gardiner, A. C. (1941). *J. Soc. Chem. Ind. Lond.*, **60**, 73.
Gibbs, R. J. (1972). *Geochim. Cosmochim. Acta*, **36**, 1061.
Graedel, T. E. (1978). "Chemical Composition of the Atmosphere". Academic Press, New York and London.
Goldberg, E. D. (1971). *Geophysics* **1**, 117.
Harvey, H. W. (1955). "The Chemistry and Fertility of Sea Water". Cambridge University Press, London.
Hem, J. D. (1970). U.S. Geol. Surv. Water-Supply Paper, No. 1473.
Hutchinson, G. E. (1975). "A Treatise on Limnology". John Wiley, New York.

Iler, R. K. (1955). "The Colloid Chemistry of Silica and Silicates". Cornell, New York.
Jones, L. H. P. (1963). *Nature, Lond.*, **198**, 852.
Jørgensen, E. G. (1955). *Physiol. Plant.*, 8, 846.
Judson, S. (1968). *Am. Scientist*, **56**, 356.
Kato, K. (1969). *Geochem. J.*, **3**, 87.
King, E. J. and Davidson, V. (1933). *Biochem. J.*, **27**, 10115.
Kharkar, D. P., Nijampurkar, V. N., and Lal, D. (1966). *Geochim. Cosmochim. Acta*, **30**, 621.
Krauskopf, K. B. (1956). *Geochim. Cosmochim. Acta*, **10**, 1.
Lagerström, G. (1959). *Acta Chem. Scand.*, **13**, 722.
Lal, D., Malhorta, P. K., and Peters, B. (1958). *J. Atmos. Terrest. Phys.*, **12**, 306.
Lisitsyn, A. P., Belyaev, Yu. I., Bogdanov, Yu. A., and Bogoyavknski, A. N. (1966). *Geokim. Kremnezema Akad. Nauk. SSR*, **37**.
Liss, P. S. (1976). *In* "Estuarine Chemistry" (J. D. Burton and P. S. Liss, eds.). Academic Press, London and New York.
Liss, P. S. and Spencer, C. P. (1970). *Geochim. Cosmochim. Acta*, **35**, 1073.
Livingston, D. A. (1963). U.S. Geol. Surv. Prof. Paper No. 440-G.
Macan, T. T. (1970). "Biological Studies of the English Lakes". Longman, London.
Maéda, H. (1952). *Publs Seto Mar. Biol. Lab.*, **2**, 249.
Maéda, H. (1953). *J. Shimonoseki Coll. Fish.*, **3**, 167.
Maéda, H. and Tsukamoto, M. (1959). *J. Shimonoseki Coll. Fish.*, **8**, 121.
Maéda, H. and Takesue, K. (1961). *Rec. Oceanogr. Wks. Japan*, **6**, 112.
Makimoto, H., Maéda, H., and Era, S. (1955). *Rec. Oceanogr. Wks. Japan*, **2**, 106.
McEwan, M. J. and Phillips, L. F. (1975). "Chemistry of the Atmosphere". Edward Arnold, London.
McRae, J. (1976). *Anal. Chem.*, **48**, 803.
Milliman, J. D. and Boyle, E. (1975). *Science*, **189**, 995.
Miyake, Y. (1965). "Elements of Geochemistry". Maruzen, Tokyo.
Moller-Anderson, J. (1974). *Arch. Hydrobiol.* **74**, 528.
Mortimer, C. H. (1941). *J. Ecol.*, **29**, 280.
Morrison, I. R. and Wilson, A. L. (1967). *Analyst*, **88**, 54.
Mullin, J. B. and Riley, J. P. (1955). *Analyt. Chim. Acta*, **12**, 162.
Narcisi, R. S. (in press). "Upper Atmosphere Physics". Gordon and Breach, London.
Pellizzari, E. D., Bunch, J. E., Berkley, R. E., and McRae, J. (1976). *Anal. Chem.*, **48**, 803.
Peterson, D. H., Conomos, T. J., Broenkow, W. W., and Scrivani, E. P. (1975). *In* "Estuarine Research". Vol. 1 (L. E. Cronin, ed.). Academic Press, New York and London.
Peterson, D. H., Festa, J. F., and Conomos, T. J. (1978). *Estuar. Coastal Mar. Sci.*, **7**, 99.
Rattray, M. and Officer, C. B. (1979). *Estuar. Coastal Mar. Sci.*, **8**, 489.
Rex, R. W. and Goldberg, E. D. (1958). *Tellus*, **10**, 153.
Riley, J. P. and Chester, R. (1971). "Introduction to Marine Chemistry". Academic Press, London and New York.
Rippey, B. H. (1980). Ph.D. Thesis, New University of Ulster.
Schink, D. R. (1967). *Geochim. Cosmochim. Acta*, **31**, 987.
Scholkovitz, E. R. (1976). *Geochim. Cosmochim. Acta*, **40**, 831.
Siever, R. (1971). *In* "Handbook of Geochemistry". Vol. 3 (K. H. Wedepohl, ed.). Springer-Verlag, Berlin.

Sillen, L. G. (1961). *In* "Oceanography" (M. Sears, ed.). Amer. Assoc. Adv. Sci., Publ. No. 67, Washington, DC.

Sugawara, K. (1967). Cited in Turekian, K. K. (1969). *In* "Handbook of Geochemistry". Vol. 1 (K. H. Wedepohl, ed.). Springer-Verlag, Berlin.

Windom, H. L. (1975). *J. Sedim. Petrol.*, **45**, 520.

Wollast, R. and DeBroeu, F. (1971). *Geochim. Cosmochim. Acta*, **35**, 613.

Yoshimura, S. (1930). *Jap. J. Geol. Geogr.*, **7**, 101.

Youngman, R. E., Johnson, D., and Farley, M. R. (1976). *Freshwater Biol.*, **6**, 253.

4

Marine Biogeochemistry of Silicon

C. P. Spencer

Marine Science Laboratories,
University College of North Wales, UK

4.1   INTRODUCTION

Silicon is the main constituent of the skeletal structures of diatoms (Bacillariophyceae), silicoflagellates (Chrysophyceae), and radiolaria (Actinopodia), and in the open oceans these are the dominant biological agencies which remove silicon from solution. In diatoms, the outer cell wall or frustule is the silicified structure and, in radiolaria, the silica skeleton is formed within the protoplasm. In both types of organism there are very great differences between different species in the robustness of the silicified structures. The diatoms are ubiquitous members of marine phytoplankton in both oceanic and neritic waters, but as predominantly obligate phototrophes their activity is necessarily limited to the upper layers of the sea. In contrast, the heterotrophic radiolaria can populate deeper waters including the abbysal regions. These protozoa are especially abundant in the warm tropical oceans but are less numerous in colder waters. Silicoflagellates (Chrysophyceae), although widely distributed, seem generally to be a minor component of the marine phytoplankton and their contribution to the biogeochemistry of silicon in sea water is therefore likely to be small.

The removal of silicon from solution by other members of the marine phytoplankton cannot however be excluded. The Prasinopyhyte, *Halosphera*, has a weakly silicified membrane surrounding the colonies which are formed by this alga (Round, 1973). A species of *Platymonas* (Prasinophyceae) accumulates considerable quantities of silicon within the cells (Chisholm *et al.*, 1978). In this latter case there is no evidence to suggest that the silicon has an essential metabolic role and its presence in the cells may be adventitious rather than functional. This also seems likely to be the status of the small quantities of silicon detected in the skeletal structures of many marine animals. The exception is found in the sponges, many species of which elaborate large numbers of siliceous spicules. These may occur as isolated bodies or be wielded together to form continuous structures. In either case they are embedded in organic tissue and their role is to increase the mechanical strength of this material. Most of the available information about the qualitative and quantitative aspects of the marine biogeochemistry of silicon relates to the effects of diatoms and to a lesser extent to radiolaria, and it is difficult to assess the magnitudes of the contributions of other biogeochemical processes to the silicon budget of the oceans. However, it seems likely that these will have only minor influences on the global oceanic biogeochemical cycling of silicon.

4.2 SILICON IN SEA WATER AND BIOGENOUS SILICA IN SEDIMENTS

4.2.1 Spatial Variation of Silicon in Sea Water

4.2.1.1 Particulate silicon

Considerable quantities of silicon are found in suspension in sea water as a major component of the particulate inorganic matter. The suspended material in sea water is often distributed very unevenly and replicate discrete samples show considerable variation (Goldberg *et al.*, 1952) but as might be expected inshore waters usually carry a heavier suspended load than water in the open ocean. The variability between discrete samples and the temporal variations in the production of biogenetic silicon in the surface layers also makes it difficult to detect systematic variations in the amounts of particulate silicon in many areas of the open ocean.

A wide variety of minerals have been recognized as components of the inorganic suspended load in sea water (Armstrong, 1965). Many of the studies of particulate aluminosilicates in sea water have been based on determinations of particulate aluminium and direct determinations of the particulate silicon are less common. A number of methods have been proposed for the determination of biogenous silicon in sediments but these do not seem to have been applied in many studies of suspended material; however, the presence of biogenous silicon may be inferred when silicon is present in excess of the requirements of the aluminosilicates, which are estimated from determinations of particulate aluminium.

Armstrong (1958) reported analyses of silicon and aluminium at three stations in the north-eastern Atlantic. The quantities of particulate silicon were in the range 19–160 µg Si l^{-1} the variability being generally less in the deeper water with values in the range 19–47 µg Si l^{-1}. The Si:Al ratios (by atoms) in many of these samples were less than 5 which suggests that much of the silicon was associated with aluminosilicates. Higher values for the ratio were recorded in the surface layers at two of the stations and at between 3000 and 4000 m at the other station where larger proportions of biogenous silicon may have been present.

Toyota and Okabe (1967) examined the particulate silicon at several stations in the north-western Pacific Ocean, the tropical Indian Ocean, and at latitudes above 60°S in the Antarctic Ocean. A range of between 0.5 and 6 µg Si l^{-1} was recorded in the open north-western Pacific south of Japan and generally higher amounts in the tropical Indian Ocean (3–17 µg Si l^{-1}). Consistently greater quantities were found in the Antarctic Ocean, most samples being in the range 10–30 µg Si l^{-1}. A number of samples had Si:Al ratios of greater than 5, most of these being recorded in the Antarctic Ocean.

Copin-Montegut and Copin-Montegut (1972) studied the suspended matter at six stations in the eastern Atlantic to the south of the stations worked by Armstrong, and they recorded generally lower quantities (range 0.5–16 µg Si l^{-1}) in the surface layers. The Si:Al ratios were sufficiently high in some samples to suggest the presence of appreciable quantities of biogenic silicon. Copin-Montegut and Copin-Montegut (1978) have also studied the suspended silicon in the Indian and Antarctic Oceans. In the tropical and sub-tropical Indian Ocean the quantities of particulate silicon in the surface layers were generally similar to those in the tropical Atlantic and although some Si:Al ratios in excess of 5 were also recorded in these waters, it is clear that aluminosilicates are usually the dominant form of particulate silicon in these areas. In contrast the total quantities of particulate silicon in the surface waters of the sub-Antarctic and Antarctic zones is much higher, the mean values of samples taken in the region of the Polar Front and parts of the Antarctic zone being in the range 35–95 µg Si l^{-1}. The Si:Al ratios of this material were in excess of 100 indicating the dominance of biogenic silicon in the suspended particulate material in these areas.

Kido and Nishimura (1975) reported similar quantities of particulate silicon in the surface water of the Japan Sea and Kuroshio area, the higher quantities (20.0 µg Si l^{-1}) in the Oyashio are being ascribed to larger diatom populations. Kido (1974) investigated the quantities of particulate silicon in the surface water of the North Pacific and recorded a systematic increase in the material at high latitudes, quantities of about 5.0 µg Si l^{-1} being typical of the water south of 30°N rising to above 20 µg Si l^{-1} at higher latitudes. There was a strong positive correlation between the quantities of particulate silicon and with both silicic acid in solution and dissolved reactive phosphate

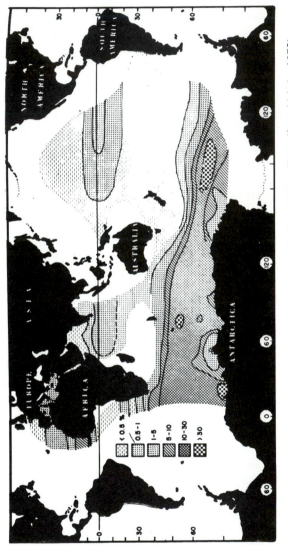

FIG 4.1(a)   Percentage of amorphous silica in suspended particulate silicon (from Lisitzin, 1972).

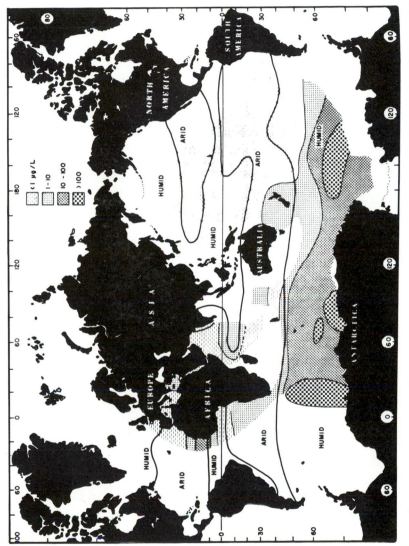

FIG. 4.1(b) Distribution of amorphous silica in suspension in the surface layer of the Oceans (from Lisitzin, 1972).

but a weak negative correlation between the particulate silicon and the particulate radioactive ^{210}Pb. If the latter constituent is used as an indicator of windborne terrestrial material the data suggests that a significant proportion of the increased quantities of particulate silicon at higher latitudes is biogenous.

There is therefore considerable indirect evidence that the biogenous particulate silicon forms an appreciable, and especially at high latitudes in the open ocean, a major proportion of the particulate silicon present in suspension. Direct measurements of amorphous silica in suspended particulate material using selective extraction into sodium carbonate have been reported by Lisitzin and others (Lisitzin et al., 1966, Lisitzin, 1970). In the Antarctic Ocean up to about 50 μg Si l^{-1} was found over a wide area with considerably greater quantities in localized regions. Similar amounts were recorded in areas of upwelling off the west coast of Africa but much of the surface water in middle latitudes carried less than 1.0 μg Si l^{-1} of amorphous silica in suspension. Although the absolute amounts of biogenic silica are low in the tropical regions the suspended material in the tropical Pacific and Indian Oceans contains a higher percentage of amorphous silica than the suspended matter in the surface layers at middle latitudes (Fig. 4.1(a)).

4.2.1.2   Silicon in solution

Removal of silicon from the surface layers of the oceans by biological action, its incorporation in skeletal structures, the sedimentation of this particulate

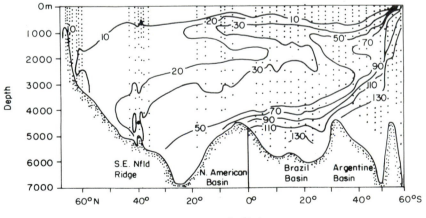

FIG. 4.2   Meridional distribution of silicic acid (μg at Si l^{-1}) in the Atlantic Ocean (from Mann et al., 1973).

silicon, and its concurrent dissolution provides a mechanism for the transport of silicon to the deeper layers of the ocean. As a consequence, the concentrations of silicon in solution in deep ocean water are always considerably greater than in the surface layers.

The concentration of silicon in solution in the deep water of the North Atlantic Ocean is about 300 µg Si l^{-1} rising to near 2000 µg Si l^{-1} at the equator and reaching over 3500 µg Si l^{-1} in the Antarctic (Fig. 4.2). A similar systematic lateral variation occurs in the deep water of the Pacific Ocean with a gradient of concentrations rising from between 3000 and 4000 µg Si l^{-1} in the Antarctic to over 5000 µg Si l^{-1} in parts of the North Pacific Ocean such as the Barents Sea (Tsunogai *et al.*, 1979). This lateral variation in the concentration of silicon in solution also occurs in the deep water of the Indian Ocean where in the North concentrations can reach between 4000 and 5000 µg Si l^{-1}. The differences in the extent of the enrichment with dissolved silicon of the deep water of the three ocean basins (Fig. 4.3) are the combined results of the silicon content of the water masses at the time of their formation and of other deep waters which may join them, and the recruitment and

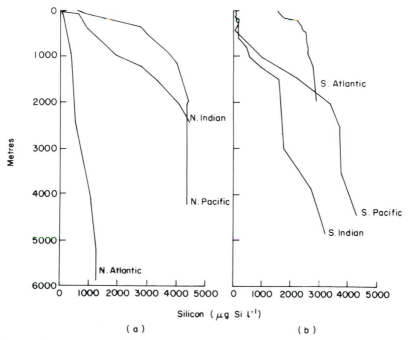

FIG. 4.3 Typical depth profiles of the concentrations of silicic acid in the Oceans (from Armstrong, 1965).

dissolution of particulate biogenous silica which continued throughout the movement of these water masses towards the south in the Atlantic Ocean and to the north in the Pacific and Indian Oceans.

Variations in the concentrations of silicon in solution and the surface waters is related to the influx of silicon from terrestrial drainage and upwelling of deep water containing high concentrations of dissolved silicon. The maximum concentrations of silicon in solution in coastal water away from the direct influence of terrestrial drainage (salinities between 34.0 and $35.0^o/_{oo}$ salinity) is often between 100 and 200 µg Si 1^{-1} but the maximum concentration in the surface layers of the open ocean is usually lower and may fall below 50 µg Si 1^{-1} in the tropics. In areas of upwelling the concentrations of silicon in solution are much higher and, for example, south of the Antarctic convergence may reach over 1000 µg Si 1^{-1}. Similar concentrations have been recorded in areas in the North Pacific.

4.2.2   Spatial Variation of Biogenous Silica in Sediments

Lisitzin (1972) has summarized the extensive studies made by himself and others on the distribution of biogenous silica in the bottom deposits of the oceans (Fig. 4.4). The Antarctic region is marked by a circumpolar band of sediments between 900 and 2000 km wide which have a high content of bio-genous silica and which, in localized regions, can reach more than 70 % of the dry weight of the sediment. A similar belt occurs in the northern Pacific Ocean but the biogenous silica content of these sediments is generally lower (10–20 %) than in the Antarctic zone although the sediments are consistently richer in biogenous silica in the Barents Sea (up to 30 %). Other areas of the northern Pacific (e.g. the Japan Sea) are ones where biogenous silica accumulates in the sediments but similar regions are absent in the equivalent northernmost areas of the Atlantic and Indian Oceans. The sediments in the Arctic Ocean normally contain about 0.5 % of biogenous silica with maximum values between 4 and 5 %. The quantities of biogenous silica are similarly low in the sediments of the Mediterranean Sea.

The sediments in the equatorial region of the Pacific Ocean contain greater quantities of biogenous silica than is characteristic of the sediments in the areas to the north and south, although the lateral variation of biogenous silica in the equatorial sediments seems to be rather irregular. A similar zone occurs in the Indian Ocean.

There are other isolated regions where sediments containing high propor-tions of biogenous silica occur as, for example, in the Gulf of California and off the west coast of Africa. Such localized accumulations are usually associated with upwelling of deep water and the sediment in these areas can contain 50–65 % of biogenous silica.

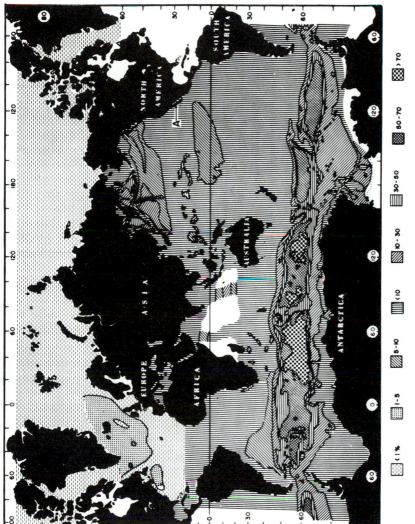

FIG. 4.4 Distribution of amorphous silica (% dry wt.) in the surface layer of sediments (from Lisitzin, 1972).

4.2.3   Temporal Variations of Dissolved Silicon in Surface Sea Water

The growth of diatoms and the concomitant removal of silicon from solution can be controlled by a large number of physical and environmental factors acting individually or in concert. It is often impossible to identify one factor as having a dominant effect in controlling the growth of populations of diatoms but in temperate latitudes at the end of the winter adequate quantities of the plant nutrients are usually present in the water and the initiation of the growth of diatoms seems to be usually controlled by a combination of physical factors. The primary requirement is sufficient radiant energy, the quantity entering the surface layers increasing systematically during this season. The effects of the increases in light on the seeding population of diatoms in the water is modified by turbidity and other factors which control the attenuation of light with depth in the water column and by turbulent vertical mixing which controls the vertical distribution of the population (Svedrup, 1953). Net growth of the population is possible when the depth of vertical mixing is less than the depth where the vertically integrated photosynthesis and respiration of the population of diatoms are equal (critical depth). These conditions are established, initially perhaps only intermittently, in temperate waters as a result of the seasonal increase in radiant energy entering the sea and a decrease in the intensity of vertical mixing. Later, in suitable conditions, a thermocline may be established which delineates the depth of vertical mixing, but this is not a necessity for net growth of the phytoplankton and in shallow water the critical depth may be greater than the total depth of the water column.

The net growth of the diatom population is accompanied by a net loss of silicon from solution and if favourable conditions are maintained the concentrations fall rapidly and can be reduced to 1 or 2 μg Si l^{-1}. The balance between uptake, grazing of the diatom stocks, concomitant regeneration, and recycling and net losses of particulate silicon as a result of sedimentation are the processes which interact to maintain low concentrations of silicon in solution in surface water.

The cycle is completed in the latter months of the year by the combination of systematic decreases in incident radiant energy and increases in the intensity of vertical mixing which together ultimately limit net growth of the phytoplankton and favour net regeneration of silicon in solution. In addition, deeper water, which has increased concentrations of silicon in solution as a result of the addition and resolution of sedimenting siliceous particulate material, may be mixed with the surface water. The sequence establishes the well-recognized annual cycle of dissolved silicon in the surface layers of the sea in temperate latitudes (Armstrong, 1965). Most of our information about such cyclical changes in temperate latitudes relates to inshore waters but it is reasonable to assume that similar changes occur in many areas of the open oceans at these latitudes, although the amplitude of the changes is probably less.

In the tropics the surface layers of the open ocean are characterized by a deep permanent thermocline so that there is a continuous net loss of particulate material from the surface layer which cannot be replenished with silicon or other plant nutrients by vertical mixing with deep water containing higher concentrations of these materials. There is also less seasonal variation in the physical factors and, except in localized regions at continental margins where mass advective upwelling of deep water occurs and at the equatorial divergencies, an internal balance between uptake and regeneration is more important than recruitment into the system. Consequently, the concentrations of silicon in solution remain low throughout the year. Menzel and Ryther (1960) have reported annual variations of between 8 and 50 µg Si l^{-1} in the Sargasso Sea. Diatoms form a proportion of the phytoplankton population in these waters but, as is perhaps generally the case in tropical waters, non-silicious species are relatively more important than in some other environments. Regular cyclical changes are much less marked in tropical waters than at other latitudes and the quantities of biogenic silicon in suspension are lower than in many other areas.

The pattern of the annual variation in incident radiant energy in polar regions allows only a relatively short period when phytoplankton growth is possible. In the Antarctic high concentrations of all the plant nutrients are available and during the Antarctic summer diatom growth removes large amounts of silicon from solution. Annual variations in concentration from about 1100 µg Si l^{-1} in winter to less than 300 µg Si l^{-1} have been recorded (Clowes, 1938). Diatoms are probably the main agency of primary production in these waters and the area is a site of formation of large quantities of biogenous silica. In contrast the Arctic Ocean is less productive although the sub-Arctic seas support larger populations of diatoms.

4.2.4 The Solubility of Silicic Acid in Sea Water

Quartz is the most stable form of silicon at natural temperatures but crystallization rates of this solid phase are exceedingly slow and the solubility of silicic acid in water is controlled by the equilibrium:

$$SiO_{2(amorphous)} + 2H_2O \rightleftharpoons Si(OH)_4$$

Values for the pK_s of the equilibrium have been reported to be in the range 2.69–2.74 at 25°C (Alexander *et al.*, 1954; Greenberg and Price, 1957; Siever, 1962). Greenberg and Price also studied the equilibrium in a range of sodium chloride and sodium sulphate concentrations and reported that the solubility was little affected in these concentrations of neutral salts. The conditional solubility constant (in 0.5 M $NaClO_4$) reported by Langerstrom (1959) also suggests that the activity coefficient of silicic acid is essentially unity up to this ionic strength. Krauskopf (1956) and Siever (1962) also reported little differences in the solubility of amorphous silicon in distilled water or sea water but Greenberg and Price recorded a pK_s of 2.79 in 1.0 M NaCl at 25°C suggesting a small decrease in solubility at higher ionic strengths.

The solubility of amorphous silica in sea water is affected by pH as a result of the dissociation of silicic acid. This effect becomes increasingly important at pH values above 9 but over the range of pH in natural sea water the effects are only slight and below pH 8 the solubility of silicic acid is essentially independent of pH (Fig. 4.5). Willey (1974) has calculated the effect of pressure on the ionization of silicic acid in sea water. At a pH of 8 about 2% is present as silicate ions and this rises to about 4% at 1000 atm.

The solubility of silicic acid in water falls with decreasing temperature. Alexander et al. (1954), Krauskopf (1956), Greenberg and Price (1957), and Siever (1962) provide solubility data at different temperatures (Fig. 4.6). The

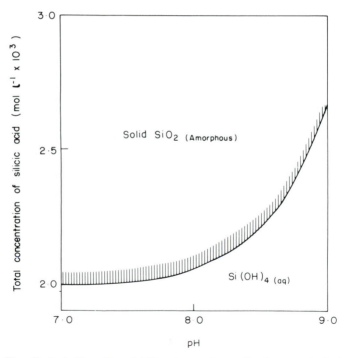

FIG. 4.5   The effect of pH on the solubility of amorphous silica in water (calculated from data in Langerstrom, 1957).

solubility of silicic acid in sea water can be expected to vary at a pressure of 1 atm. from about 28 000 μg Si l^{-1} at 0°C to 69 000 μg Si l^{-1} at 25°C.

The effect of pressure on the solubility of amorphous silica has been investigated by Jones and Pythowicz (1973) and Willey (1974). The latter worker demonstrated than an increase in pressure from 1 to 1220 atm. caused an increase in the solubility in sea water of 0°C of about 43 %. The relationship between the solubility and pressure includes two distinct phases (Fig. 4.7). These are compatible with differences in the net change in partial molar volumes which occur during the solution process and suggest values for $\Delta \bar{V}$ of -18 cm^3 mol^{-1} up to pressures of 200 atm. and -8 cm^3 mol^{-1} between 200 and 1200 atm. It is possible that these differences reflect changes in the number of hydrating water molecules per solute molecule. It is therefore clear that sea water throughout the oceans is undersaturated with respect to amorphous silica. The available physico-chemical data and exerimental evidence indicate that monomeric silicic acid is the predominant species of silicon present in solution in sea water and the form in which it

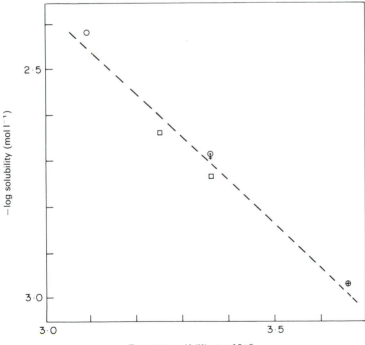

FIG. 4.6   The effect of temperature on the solubility of amorphous silica in water (from data in Alexander *et al.*, 1954 (○); Greenberg and Price, 1959 (□); Krauskopf, 1956 (+); and Siever, 1962 (●)).

is removed from solution by biological activity (see Chapter 3). In contrast to the biological removal of calcium, which is present in supersaturated solution in much of the surface water of the oceans, the uptake of silicon occurs from solutions which are undersaturated with respect to amorphous silica.

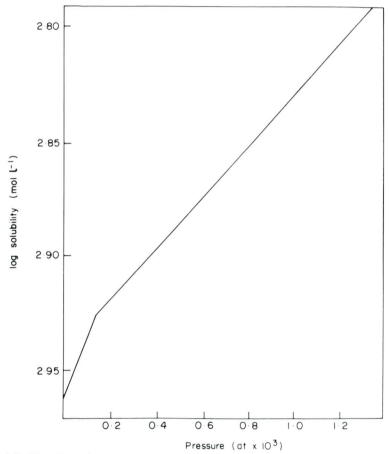

FIG. 4.7   The effect of pressure on the solubility of amorphous silica in water (from data in Willey, 1974).

4.3   THE REMOVAL OF SILICIC ACID FROM SOLUTION IN THE SEA BY DIATOMS

4.3.1   Silicon Content of Diatoms

There are well-recognized differences in the silicon content of diatoms. The thickness of the silicified cell wall varies considerably from one species to another and, in addition, the size of the cell walls of a population of centric diatoms changes during the growth of the population. The degree of silicification of the cell walls in some species varies considerably if the supply of silicic acid in solution is exhausted (Paasche, 1973a) and in such species cell division can continue for a period with the development of increasingly thinner cell walls. Other species are apparently less adaptable and cell division stops when the supply of silicic acid is exhausted.

Variations in the silicon content of different species of diatoms was evident in the results of the early analyses of marine diatoms. Parsons et al. (1961), using cultures of Skeletonema costatum and a species of Coscinodiscus, measured Si:C ratios (by atoms) of 0.19 and 0.43 respectively. Vinogradov (1953) had earlier reported a range of from 0.17 to 0.59 for the ratio. McAllister et al. (1961) and Antia et al. (1963) working with a mixed pipulation of coastal phytoplankton (mainly diatoms) reported ratios between 0.15 and 0.41 during the growth of the populations and these results allowed Strickland (1965) to suggest that a Si:C ratio of near 0.34 was probably appropriate in areas where silicic acid was not limiting growth.

Copin-Montegut and Copin-Montegut (1978) were able to make estimates of the Si:C ratio in the natural particulate material present in the oligotrophic subtropical Indian Ocean. In the area where aluminosilicates are an important fraction of the particulate silicon, multiple regressions of the data suggested a Si:Al ratio of near 2 and a biogenic Si:C ratio of 0.015 (by atoms) which was within the range for this area reported by Lisitzin (1967). Further south, as noted earlier, the relative quantities of aluminium present in the particulate material are very much lower and most of the particulate organic carbon is associated with the populations of diatoms. Si:Al ratio in these waters exceeded 100 in some areas and the mean Si:C ratio was about 0.4. Higher values were reported by Lisitzin in these areas but differences in methodology may be a cause of these discrepancies. The Si:C ratio of natural particulate matter will be expected to show considerable variation not only because of differences in the degree of silicification of the cells in populations which are floristically different as noted above but also as a result of geographical and temporal variation in the relative proportions of diatoms and non-siliceous species in the standing crops of phytoplankton.

Later experimental studies by Eppley et al. (1967) and Strickland et al. (1967) using the diatom Ditylum brightwelli showed Si:C ratios (by atoms) of 0.28 to 0.58. In experimental studies using greater control of the nutrient status of the cells, Durbin (1977) recorded Si:C ratios of 0.37 to 0.41 in Thalassiosira nordenskioeldii. Harrison et al. (1977) reported a ratio of 0.11 in Skeletonema costatum the growth of which was not limited by silicic acid, nitrogen or phosphorus, and values of 0.17 and 0.16 for Chaetoceros debilis and Thalassiosira gravida under the same conditions. If the growth of these diatoms was limited by a shortage of silicic acid in the medium the Si:C ratios fell to 0.03 to 0.04. In contrast, limitation of the growth rate of the diatom by the nitrogen source (ammonium-nitrogen) resulted in the Si:C ratios in the cell rising to between 0.3 and 0.4 in the case of *Skeletonoma costatum* and *Chaetoceros debilis* but the silicon content of the cells of *Thalassiosira gravida* was little affected at reduced growth rates caused by suboptimal concentrations of ammonium-nitrogen.

The ratios in which silicon, nitrogen, and phosphorus (ΔSi:ΔN:ΔP) appear in some deep water as a result of regeneration of the organic material produced by diatoms has been reported to be 16:16:1 in the Atlantic Ocean (Richards, 1958) and 16:14.3:1 in the Irminger Sea (Stefasson, 1968). The apparent utilization of oxygen associated with these ratios suggests a Si:C ratio of between 0.13 and 0.14 (by atoms) in the original cellular material and a lower value of about 0.07 is implied by similar data from samples for the North Sea (Schott and Ehrhardt, 1969). Park (1967) reported that the ratio of apparent oxygen utilization to silicon in samples of deep water taken off the Oregon coast agreed well with values reported by Grill and Richards (1964) during laboratory studies and these implied a Si:C ratio of 0.21. The latter studies were obtained with material which consisted mainly of a species of *Coscinodiscus*, most species of which are relatively heavily silicified. Values of the Si:C ratio obtained in this way must be minimal values because a proportion of the dissolution of biogenous silica occurs after the oxidative phase of regeneration has been completed (Section 4.5.2).

There is little information about the inorganic constituents of diatom cell walls other than silicon. The analyses reported by Parsons *et al.* (1961) showed that there was little difference in the percentage silicon in the ash weight of *Skeletonema costatum* and *Cosinodiscus* sp. which usually have cell walls of very different thickness. This suggests no major differences in the inorganic composition of the cells walls of these two species. Werner (1978) has determined the elementary composition of intact and plasmolysed cells of the diatom *Cosinodiscus asteromphalus*. His results show considerable enrichment of the treated cell walls with sulphur, zinc, and copper. The plasmolysis only removed 84 % of the organic matter and it is impossible to decide if the zinc and copper retained in the treated material was associated with the siliceous cell wall or was complexed with the residual denatured organic matter. There are obviously considerable practical difficulties in the investigation of the elementary composition of the cell walls themselves because any treatment of intact cells which efficiently removes the cellular organic matter is likely to remove trace constituents from the cell walls, and there is also the possibility of subsequent adsorption of other material by the siliceous structure.

However, because of the adsorptive properties of the surface of hydrated silica, it would be surprising if the siliceous cell walls of diatoms did not include some trace metals. Adsorption will also be favoured by the very large surface area of many diatoms as a result of the complex morphology of the structure of the cell walls. Werner (1978) cites the case of the cells of *Coscinodiscus asteromphalus* of which the surface area is 100 times greater than a simple cylinder of similar dimensions.

Hurd (1973) reports an analysis of biogenic silica in a sediment from the Pacific which presumably mainly consisted of the remains of radiolarian skeletons. This material was cleaned by an ultrasonic treatment and contained 1700 ppm iron, 770 ppm aluminium, 270 ppm calcium, 205 ppm chromium, 54 ppm nickel, and 36 ppm cobalt. At least some of these constituents may have been acquired during the sedimentation through the water column of the skeletons or during their residence in the sediment.

The silicon in the cell wall of diatoms seems to be predominantly in the form of amorphous silica or opal (Kamatini, 1971). Small amounts (1 %) of two types of strained quartz have also been detected in the cell walls of freshly killed diatom (Gregg *et al.*, 1977).

4.3.2 The Incorporation of Silicic Acid in the Cells Walls of Diatoms

Studies using cultures of diatoms show that silicic acid in solution is an absolute requirement for the reproduction and growth of these algae. If adequate concentrations of silicic acid are available in solution, the growth rate of diatoms and the rates of uptake of silicic acid are independent of its concentration. Below a certain concentration, the growth rate and rate of removal of silicic acid from solution decrease with decreasing concentrations. Growth and, hence, uptake of silicic acid are prevented at some low but finite concentrations of silicic acid. The possibility that the growth and primary production by diatoms in nature might in some areas be limited by the supply of silicon has stimulated a number of studies on the kinetics of these processes. Growth and nutrient kinetics have been studied using pure cultures of individual diatoms under non-steady state conditions (Paasche, 1973a; Guilliard *et al.*, 1973; Thomas, 1975; Davies, 1976), after transient perturbation to a steady state (Davies, 1976; Conway and Harrison, 1977), under steady-state conditions in chemostats (Paasche, 1973b; Harrison *et al.*, 1976; Davies, 1976) and using isotopic tracer with pure cultures and natural populations (Goering *et al.*, 1973; Nelson *et al.*, 1976).

The relationship between the growth rate of an alga and the concentration of a nutrient which is controlling the growth rate is non-linear. Earlier work assumed that a rectangular hyperbola of the type which describes typical Michaelis-Menton kinetics such as exhibited by enzymatic reactions also adequately described algal growth kinetics. However, more recent work has shown that kinetic data of this sort does not always fit a hyperbola which passes through the origin of a plot of growth rate against the concentration of silicic acid. A satisfactory fit can sometimes be obtained using a hyperbola of the form:

$$\mu = \mu_{max} \frac{Q - Q_0}{K_s + (Q - Q_0)}$$

where μ_{max} is the maximum growth rate when the silicic acid is present in excess of growth rate limiting quantities, μ is the growth rate where the cell quota of silicon is Q, Q_0 is the minimal cell quota of silicon below which growth is inhibited, and K_s is a constant.

When diatoms are grown in a chemostat at a constant flow rate with silicic acid as the limiting nutrient and steady-state conditions are established in the reaction vessel, the growth rate of the alga must equal the dilution rate, and measurements of the steady-state concentrations of silicic acid make it possible to determine μ_{max} and K_s. The results obtained by Paasche (1973b) with the diatom *Thalassiosira pseudoana* suggested that the alga could not reduce the concentration of silicic acid in solution below about 19 µg Si l^{-1} and gave values of $(K_s + Q_0)$ of between 34 and 53 µg Si l^{-1}. Harrison *et al.* (1976) using the diatom *Skeletonema costatum* investigated the relationship between the dilution rate (D) and the steady state concentration of silicic acid but their data did not fit a simple hyperbola or, except over a limited range of dilution rate, the relationship proposed by Droop (1973):

$$D = D_{max} \left[1 - \frac{k_Q}{Q} \right]$$

where D is the dilution rate when the cell quota of silicon is Q, D_{max} is dilution rate approached when $Q \gg k_Q$ and k_Q is a constant which may be thought of as the cellular subsistance quota of silicon. However, these results confirmed Paasche's observation that this diatom was unable to reduce the concentration of silicic acid below a minimum value of between 14 and 28 µg Si l^{-1}.

Studies made under non-steady state conditions may be complicated by the common observation that the growth rate of algae do not depend on the immediate nutrient conditions but on those which prevailed during the preceding period. Furthermore, in some studies the average growth rates over a period are related to the initial concentration of a nutrient which is in fact changing continuously. Paasche (1973a) working with *Skeletonema costatum*, *Thalassiosira pseudonana*, *Thalassiosira decipiens*, *Ditylum brightwellii*, and a species of *Licmophora* showed that growth ceased in batch cultures when the concentration of silicic acid was between 14 and

$42 \ \mu g \ Si \ l^{-1}$. Paasche studied the kinetics of uptake of silica deficient cells of these five species of diatoms and was able to fit the data to the expression:

$$V = V_{max} \left[\frac{(S - S_0)}{K_v + (S - S_0)} \right]$$

where V is the velocity of uptake at a silicic acid concentration S, V_{max} is the maximum uptake rate at very high values of S, S_0 is the silicic acid concentration below which there is no net uptake, and K_v is a constant. The values of S_0 were estimated to be in the range $9-37 \ \mu g \ Si \ l^{-1}$ and mean values of $(K_v + S_0)$ in the range $31-132 \ \mu g \ Si \ l^{-1}$. In other studies under non-steady state conditions Guillard et al. obtained values of K_s of 5.3 and $27 \ \mu g \ Si \ l^{-1}$ for their strains of Thalassiosira pseudonana, and Thomas (1975) reported values of K_s for Chaetoceras gracilis of 13 and of 9 and $21 \ \mu g \ Si \ l^{-1}$ for two species of Leptocylindricus.

Davies (1976) produced silicon deficient cells of Skeletonema costatum in a chemostat and then studied the rate of uptake following an addition of silicic acid. His data showed that the rate of uptake of silicic acid was related to a simple hyperbola in a limited range of silicic acid concentrations with values of K_v of between 31 and $50 \ \mu g \ Si \ l^{-1}$. In similar, but not identical studies, Harrison et al. (1976) and Conway and Harrison (1977) made a detailed examination of the kinetics of uptake by three diatoms following an addition of silicic acid to cells grown under silicon limited conditions in a chemostat. The three species differed in their behaviour but in each case one phase of the uptake process which included silicic acid concentrations up to $168 \ \mu g \ Si \ l^{-1}$ (Chactoceros dedilis), $84 \ \mu g \ Si \ l^{-1}$ (Skeletonema costatum), and $42 \ \mu g \ Si \ l^{-1}$ (Thalassiosira gravida) showed suitable kinetics for fitting to a hyperbola. Values of K_v for the three species were calculated to be 70, 42, and $22 \ \mu g \ Si \ l^{-1}$ respectively and concentrations of silicic acid of 8.4, 5.6, and $14 \ \mu g \ Si \ l^{-1}$ respectively were the minimum to which these species were able to reduce the silicic acid in solution.

Goering et al. (1973) used stable isotopes (^{29}Si and ^{30}Si) for measuring the kinetics of uptake of silicic acid by natural populations of phytoplankton. The value of K_v for these populations was calculated to be $82 \ \mu g \ Si \ l^{-1}$. More recently Nelson et al. (1976) have used the same techniques to investigate silicon uptake by the two strains of Thalassiosira pseudonona studied by Guillard et al. (1973). Values for K_v for the two clones did not show as large a difference and were considerably higher than K_s values reported by Guillard et al. (1973). Nelson et al. (1976) were also able to measure the concurrent rates of silica dissolution which ranged from 6.5 to 15 % of the uptake rates; however, these rates were insufficient to explain zero rates of net uptake at low but finite concentrations of silicic acid and it is possible that the modified hyperbola used by Paasche (1973a) describes the kinetics of uptake of silicic acid by this population better than the hyperbola passing through the origin used by Nelson *et al.*

The experimental data summarized above make it certain that considerable differences occur in the ability of different species of diatoms to remove silicic acid from solution and also between different strains of the same species. The units of the constants K_s and K_v are concentrations and if the relationship between the rates of growth or uptake are adequately described by a rectangular hyperbola the constant is numerically equal to the concentration of silicic acid which allows growth or uptake to proceed at half the maximum rate.

It is difficult, in the case of those data which necessarily relate to artificial conditions, to apply them to natural populations of diatoms. In the sea the uptake of silicon may be affected by interactions with other nutrients which may override control of silicon uptake by the ambient concentrations of silicic acid and light intensity has considerable effect on the kinetics of silicon uptake (Nelson *et al.*, 1976; Davies, 1976). The differences reported in the magnitude of the constants which characterize the kinetics of growth or uptake are clearly often the result of the complicated kinetics of uptake in non-steady state conditions and of variations in the metabolic control of the growth rate at different growth rates in steady-state conditions in chemostats.

However, it seems probable that of the limited number of diatoms investigated that variations in their ability to remove silicic acid from solution may have ecological significance. It is possible that the low concentrations of silicon which occur at times in surface water in many parts of the oceans are in the range where the growth rate of diatoms can be related to the ambient concentration of silicic acid, and that under some conditions the quantity of silicic acid might act as a controlling factor for the growth of diatoms. Moreover, it seems certain that some diatoms at least have the potential to strip silicic acid sufficiently from solution in sea water to maintain the very low concentrations observed on some occasions. On these grounds there are no reasons for postulating the operation of other abiological processes which remove silicic acid from solution in the surface waters of the open ocean.

4.3.3   Aspects of the Chemistry of the Formation of the Siliceous Cell Walls of Diatoms

The shell of diatoms also includes some organic matter. Studies using electron microscopy (Schmid and Schulz, 1979) demonstrated that the siliceous structure is laid down from vesicles which fuse with a membrane and release the contents of the vesicle on to the deposition sites. Heckey *et al.*

(1973) investigated the amino acid and carbohydrate composition of the cell walls of several diatoms which had been cleaned ultrasonically. The spectrum of amino acids associated with the cell walls was considerably enriched with the hydroxyl amino acids serine and threonine, and with glycine. The carbohydrate fraction was more variable in composition and the individual sugars in this polysaccharide therefore seem to have a less specific role than the amino acids.

Heckey *et al.* (1973) suggest that the hydroxyl groups of the serine and threonine may provide suitable sites for condensation reactions with silicic acid and provide a suitable template. Successive accumulations of a hydrated silicon matrix as a result of further polycondensation reactions may be favoured by the orientation of the hydrated silicon units which first condense on the amino acid surface. The polysaccharide component which it is suggested lies outside the amino acid layer seems less likely to be directly involved in the accumulation of silicon, and it may have a protective role.

Although the accumulation of silicon by diatoms is inhibited by dinitrophenol, the total energy requirements for the process is low in comparison with the requirements for the synthesis of the typical structural polysaccharides of many algae. Werner (1977) has pointed out that the energetically favourable elaboration of siliceous structural material may provide diatoms with an ecologically significant advantage over other unicellular algae.

The relationship between the processes involved in the accumulation of silicon by diatoms and other metabolic activities of the cell is very close. If diatoms are transferred to a medium which contains no silicic acid, cell division is inhibited in some species within 1 h, and further protein and DNA synthesis is stopped after 4 h (Werner, 1977; Darley and Volcani, 1969), although as noted earlier there is some variability between different species in this respect and the time scale of these metabolic effects can be delayed if the species is one which is able to adapt by forming an increasingly thinner cell wall. Inhibition of chlorophyll and carotenoids synthesis follows with a subsequent sharp drop in photosynthetic activity. The biochemical effects of silicon deficiency therefore manifest themselves in effects on a number of metabolic systems as well as in the elaboration of the structural material of the cell wall.

The detailed mechanisms of these interactions are unknown. There is evidence that silicic acid reacts with a number of physiologically active materials such as amino acids. The demonstration that boron is an essential element for diatoms (Lewin, 1965, 1966) and the growth-promoting activity to some diatoms of silica borates suggests an important role for these compounds, but their function is not known.

The presence in the cytoplasm of diatoms of a water soluble silicic acid fraction associated with a constant amount of protein has been demonstrated

by Werner (1977) and Mehard *et al.* (1974). This fraction may act as the
internal pool of silicon which controls growth. It has been noted in Section
4.3.2 that such an internal pool or quota has been postulated to provide a
more satisfactory model of the kinetics of silicon uptake and growth.

4.4   THE DISSOLUTION OF SILICA IN SEA WATER

4.4.1   Factors Affecting the Rate of the Dissolution

The rates of dissolution of various forms of silica in sodium hydroxide were
investigated by Greenberg (1957). The rates vary considerably from one form
of silica to another, the rate of dissolution of quartz being very slow at the
natural temperature of sea water. The hydrolysis of polycondensation polymers often follows second-order kinetics, but the hydrolysis of the Si–O–Si
bond is apparently more complex and the order of the reaction may
change during the dissolution process. Kato and Kitano (1968) also recorded
complex kinetics during the dissolution of amorphous silica in distilled water
and sea water. The rate of dissolution of silica is a function of temperature,
the degree of crystallinity of the solid material, and its surface area and can be
altered by mechanical or heat treatment.

The hydrolysis of the Si–O–Si bond is influenced by hydroxyl ions and
Krauslopf (1959) and Kato and Kitano (1968) reported that the rates of
dissolution of amorphous silica were higher in sea water (pH 8.2) than in the
less alkaline distilled water. Greenberg (1957) reported that the rates were
unaffected by addition of sodium sulphate and no effect of increases in ionic
strength is therefore to be expected, although no studies appear to have been
made in sea water.

The dissolution process can be represented by:

$$SiO_{2(s)} + 2H_2O \rightleftharpoons Si(OH)_{4(aq)}$$

and O'Connor and Greenberg (1958) described the processes by the
expression:

$$\frac{dC}{dt} = k_1 S - k_2 C \cdot S$$

where C is the concentration of silicic acid in solution, S is the surface area of
the solid phase material and k_1 and k_2 are the rate constants for the
dissolution and polymerization reactions. Using an integrated form of the
expression:

$$\log \left[\frac{C_s - C}{C_s} \right] = k_2^* S \cdot t$$

where C_s is the concentration of silicic acid in a saturated solution and $k_2^* = -k_2$, they were able to obtain a good fit of experimental data.

The solubility of the amorphous silica used by O'Connor and Greenberg (standard luminescent silica) varied with the quantity of the solid phase material which suggests that the material was not homogenous.

4.4.2 The Dissolution of Biogenous Silica

Jørgensen (1955) investigated the rates of solution of the cell walls of two species of diatoms. The cells were killed by treatment with methyl alcohol and suspended in bicarbonate-carbonate buffers. Little solution occurred at pH 5.0 but increases in the rate of solution with increasing pH were evident. The cell walls of the two species behaved differently which suggested differences in the nature of the silica.

Lewin (1961) obtained similar results with a different species and she confirmed that the dissolution in Tris buffer of silica from the cell walls of untreated diatom was slow compared with the behaviour of cells which had been killed and treated in various ways and it appeared that removal of metal ions from the cell walls assisted dissolution. Lewin also showed that the dissolution of the silica in the cells walls of natural populations behaved similarly when suspended in buffer or sea water. The rates of dissolution of fossil diatom cell walls were only slightly increased by the treatments which were effective with fresh material.

During an investigation of the regeneration in sea water of nutrients from a sample of marine phytoplankton which consisted largely of an unidentified species of a centric diatom, Grill and Richards (1964) also recorded the release of silicic acid. Virtually all the particulate silicon disappeared in 126 days and the dissolution process followed first order kinetics with respect to the particulate silicon with a rate constant of $0.017 \, \mathrm{d}^{-1}$ (temperature not given). The concentration of silicic acid in solution only reached about 3% of the likely saturation value and under these conditions the process may approximate to first-order kinetics.

The amorphous silica in the cell walls of living diatoms is clearly stabilized against dissolution but Lewin's results suggest that the polysaccharide and amino acid components of the cell wall are not of major importance to this. On the other hand there is a considerable amount of evidence that killing diatoms causes an immediate increase in the dissolution of silica. The kinetic studies described in Section 3.2 suggest that there is some dissolution of the silica from living diatoms growing in rate controlling concentrations of silicic acid. Nelson and Goering (1977) have also presented evidence that a similar process also occurred with a natural population of marine diatoms. These observations suggest that the amorphous silica in the cell walls of diatoms may be in dynamic equilibrium with the silicic acid in solution in the medium, perhaps with the involvement of the soluble silicon-protein complex as an intermediate. Unless the intracellular pool of silicon is maintained by metabolic processes the rates of polymerization of the silicic acid may fall below the rate of spontaneous dissolution.

The observation by Jørgensen (1955) and Kamatani (1969) of differences in the rates of dissolution of silica from the cell walls of different species of diatoms may be the result of differences in surface area and the same factor may, in part, be the cause of the low rates of dissolution exhibited by fossil diatoms. However, Lewin (1961) was unable to detect any effects on the rates of dissolution of the cell walls of *Navicula pelliculose* as a result of increases in surface area following disruption into fragments of the majority of cells in an experimental sample. There is some evidence of differences in the crystallinity of the cell walls of different species of diatoms (Kamatani, 1971) and although recent diatom deposits seem to be mainly amorphous, geologically older material shows different X-ray diffraction patterns. Gregg *et al.* (1977) have demonstrated that the crystallinity of the silica in the cell walls of a freshly killed diatom changes during storage of the material for a year at room temperature. It seems most likely that it is changes such as these which decrease the rates of dissolution of the cell walls of these algae over a length of time and which are important in modifying the dissolution rates of the silica in diatomaceous sedimentary material. In this latter case the stabilizing role of iron or aluminium incorporated in the silica may also be important, as the effects of treatment with chelating agents show.

Hurd and Theyer (1975) have pointed out that amorphous silica is metastable under the conditions which occur in sediments and that there is a clear trend of decreasing solubility with increasing age of such material. This can result in a hundredfold decrease in the dissolution rate. The progress of decreasing solubility suggests that two separate processes are operating, one leading to a crystabolite structure and the other to a quartz structure. Decreases in the surface area of the particles of amorphous silica must have some influence but both morphological and diagenic changes seem likely to be involved.

The rates of dissolution of radiolarian skeletons after exposure for 4 months at various depths have been investigated by Berger (1968). The material was acid washed and treated with hydrogen peroxide. The highest rates were recorded in the surface water and below about 3000 m the rates were uniform.

Hurd (1972) has determined the rate constant for the dissolution of biogenic amorphous silica from sediments in the central equitorial Pacific. The material was cleaned by acid washing and treatment with ultra-sonics.

He determined rate constants defined by:

$$\frac{dC}{dt} = K_2(C_s - C)S$$

and recorded mean values of 1.36×10^{-7} and 2.1×10^{-8} cm s^{-1} at 25° and 3°C respectively. In a later paper (Hurd, 1973) he reported a direct linear relationship between the value of the rate constant of acid cleaned biogenic silica in sea water and pH. The rate constant recorded with the material at 3°C and pH 7.6 (about 7×10^{-9} cm s^{-1}) where slightly lower than in the earlier work.

The rate constant reported by Grill and Richards (1964) for the regeneration of silicic acid during the biological oxidation of the organic matter in the cells of a diatom is similar to the value for 25°C recorded by Hurd (1972). Apart from this work, there does not seem to be any experimental data which allows the rate constant for the dissolution of the silica in the cell walls of diatoms to be estimated.

Observations which allow estimation of the rate constants of silica dissolved in nature have been made by several workers. Berger's (1968) data suggest values for the dissolution rates of radiolarian skeletons in deep water which are similar to those reported by Hurd (1972).

Hurd (1973) calculated the dissolution constant of silicic acid (k_2) in sediments of the equitorial Pacific Ocean using the expression:

$$\frac{C_z - C_F}{C_o - C_F} = \exp Z(k_2S/D)^{1/2}$$

where C_o, C_z, and C_F are the concentrations of silicic acid in the pore water at the interface between the sediment and the water, at the bottom of the core when the concentration ceases to increase with depth, and at a depth Z respectively; S is the surface area of the biogenic silica and D the diffusion coefficient of silicic acid in solution in the sediment corrected for the tortuousity of the diffusion path caused by the presence of the solid particles. The expression assumes that the rate of diffusion of silicic acid from the sediments is controlled only by the rate of solution and rate of diffusion. The equation was fitted to observed profiles of silicic acid and the appropriate value of k_2S, and hence k_2, calculated. The values of the k_2 generated varied between 1×10^{-13} and 1×10^{-11}. These values are considerably lower than those measured with cleaned biogenic silica from the same sediments and although there are a number of approximations and uncertainties in the application of the model for the determination of the rate constants, the discrepancies are too large to be the result of these alone.

The experimental determination of the rate constants which return the highest values have generally been made with material which has been cleaned by washing with acid and other treatments (see Section 4.5.2.2). Such treatment was shown by Lewin (1961) to increase solution of the silica in the cell wall of diatoms, and it seems most probable that the higher rates recorded in these experimental studies are the result of the removal of material which stabilizes the solid phase biogenic silica.

On the other hand, Grill and Richards (1964) recorded rates of silicic acid regeneration using untreated diatoms which seem comparable to the rates measured with cleaned material. The solubilization of silica by bacterial action has been demonstrated by Laumers and Heinen (1974) and Purushothamam *et al.* (1974) have demonstrated the presence in an estuarine sediment of bacteria which solubilized magnesium trisilicate. Other micro-organisms also have this ability (Henderson and Duff, 1963). Microbiological action may therefore have accelerated the dissolution of silicic acid during the process investigated by Grill and Richards. Bacterial activity of this type must depend on an adequate supply of organic matter and the process does not seem likely to be important in oceanic sediments. The mechanism of solubilization of silica by bacteria has not been investigated. It has been suggested that this solubilization of silica is always associated with pro-teolytic activity which can release basic amines. In a micro environment this might cause a rise in pH sufficient to increase the dissolution process. Hurd (1973) reported a linear increase in the rate constant for the dissolution of biogenic silica up to pH 8.7 in sea water. The rates of dissolution of sponge spicules in reef sediments are also seemingly high and Land (1976) has suggested that the holes which appear in the spicules in this environment may be the result of attack by bacteria. The process may be important in the dissolution of biogenic silica in the upper part of the water column where the detrital remains of the cell walls of diatoms must often be included in faecal pellets or associated with other particulate organic matter which will provide a favourable environment for such bacterial action.

4.5   THE FLUXES OF BIOGENOUS SILICA IN THE OCEANS

4.5.1   Sources of Biogenous Silica

The global production of biogenous silica by phytoplankton in the oceans must be related to the global rates of photosynthesis. There have been a number of estimates of the annual production of organic carbon by the marine phytoplankton, and the more recent estimates have had the advantage of the increased amount of data which has been available as a result of the widespread use of the ^{14}C technique. The interpretation of these measurements, and in particular their extrapolation to the rates in nature, present many difficulties. The estimates of global photosynthesis which have been made differ mainly as a result of the method of sub-division of the oceans into areas of different productivity and assumptions in the weighting given to various areas, a particular difficulty being the uncertainty about the extent of the area of greatest production in the Antarctic Ocean where there seems to be considerable lateral variations. Koblentz-Mishke *et al.* (1970) calculated a value of 2.3×10^{16} g C y^{-1} and this figure was raised to 3.3×10^{16} g C y^{-1} by Bruyevich and Ivanenkov (1971). Platt and Subba Rao (1973) calculated an annual rate of 3.1×10^{16} g C y^{-1} and at present therefore a value of about 3.0×10^{16} g C y^{-1} is probably the best estimate available.

Apart from the uncertainties introduced in extrapolation of results from the ^{14}C technique to estimate total photosynthesis in the water column, there is considerable doubt about the relationship between the rates of retention of ^{14}C by algal cells during such measurements and the true rates of photosynthesis. The new cellular carbon which appears in suspension (net photosynthesis) is resultant of the total fixation of carbon dioxide (gross photosynthesis) and the concurrent losses as a result of respiration and by such processes as the loss of soluble organic matter from the cells. The extent of the latter process in nature is uncertain but it may sometimes represent a significant proportion of the gross photosynthetic production of organic carbon by phytoplankton.

Losses of organic carbon by respiration or in any organic matter released will not be recorded by the ^{14}C technique, but it cannot be assumed that the method provides a reliable measurement of net photosynthesis (i.e. gross photosynthesis − respiratory losses). It is often considered that measurements made using the ^{14}C technique represent something between gross and net photosynthetic rates. These considerations may be important in the estimation of the global rates of carbon dioxide fixation and it is arguable that the estimates based on measurements of ^{14}C retention may be biased towards a minimum value. On the other hand, for estimates of the production of biogenic silica by diatoms the value of the net photosynthesis is more directly relevant.

Another considerable difficulty in the estimation of the global rate of formation of biogenic silica by phytoplankton is the big variation in the contribution of siliceous species to the total photosynthetic activity in different areas. It is well recognized that non-siliceous species, many of which are very small and escape detection by some techniques for enumerating phytoplankton population, are very important contributors to the annual photosynthesis in many areas. Although the total biomass of these organisms may often not be large, there is evidence that their photosynthetic activity is sometimes disproportionately high. The relationship between photosynthesis and the formation of biogenic silica is therefore certain to vary from area to area.

Platt and Subba Rao (1973) suggest that the total production of the oceanic offshore areas amounts to 23.6×10^{15} g C y^{-1}. This production is distributed between three types of areas each with a different mean annual rate of photosynthesis. Allocating the total production in the ratios of the mean rates suggested by Koblentz-Mishke *et al.* (1970) and weighting the contribution of oligotrophic, transitional, and grouping together oceanic areas of divergencies and the sub-polar open ocean according to their areas, suggests annual rates of photosynthesis of about 4.1, 8.1, and 11.4 $\times$ 10^{15} g C y^{-1} respectively in these environments.

The proportions of this photosynthesis due to diatoms in the three types of areas are more difficult to estimate as a result of major systematic floristic differences in the phytoplankton populations and in seasonal variations in production in temperate latitudes. In the oligotrophic areas the production of organic carbon may depend on more efficient cycling of phosphorus and nitrogen in the surface layers as well as the influence of nitrogen fixation. Such an ecosystem will discriminate against diatoms and as little as 10% of the photosynthesis in these areas may be accompanied by the formation of biogenic silica. The surface waters of the transitional areas seem to carry quite low loads of amorphous silica in suspension (Lisitzin, 1972) and making an allowance for seasonal variation it may be reasonable to adopt a value of 25% of the contribution by diatoms in these areas. In oceanic regions of divergence and in sub-polar waters of higher productivity the populations of diatoms are larger and perhaps contribute 50% of the annual photosynthesis. These estimates yield values of 0.4, 2.0, and 5.7 $\times$ 10^{15} g C y^{-1} for the contributions of diatoms to the annual rates of photosynthesis in oligotrophic, transitional and more fertile oceanic areas such as areas of divergence and sub-polar waters.

The areas classed by Platt and Subba Rao (1973) as shelf areas have a total production of 7.3×10^{15} g C y^{-1} of which 45% is accounted for by Antarctic production and the remainder by areas of upwelling and fertile neritic regions. Assuming that diatoms are responsible for all the production in Antarctica and making some allowances for the seasonal variation of their contribution by allocating a proportion of 75% of the production in the remaining shelf areas, this suggests that 6.3×10^{15} g C y^{-1} are produced by diatoms in these areas. The total contribution of diatoms to the global rates of photosynthesis may therefore be as little as 14.0×10^{15} g C y^{-1} or only about 45% of the total.

The conversion of these rates of photosynthesis by diatoms to rates of production of amorphous silica rely on the use of Si:C ratios. The variability of these has been discussed in Section 4.3.1. The degree of silification in diatom populations seems likely to be related to the availability of silicic acid in solution and it may be expected that populations growing in high concentrations of silicic acid will contain a high proportion of species with heavy cell walls. In contrast, in oligotrophic areas lightly silicified species are likely to be dominant. The variations in the Si:C ratio of natural suspended material reported by Copin-Montegut and Copin-Montegut (1978) and Lisitzin (1966) are probably in part at least a reflection of these systematic floristic differences. Estimation of the total annual rates of formation of biogenous silica using a constant Si:C ratio is not very realistic and use of a ratio as high as 0.46 (2.3, SiO_2:C by weight) seems likely to result in high estimates. In the light of the present information about the value of the ratio recorded in Section 4.3.1 it seems reasonable to use ratios varying from 0.015 to 0.4 for the different domains in which the annual production of organic carbon has been allocated above.

An estimate of the annual global production of amorphous silica by diatoms is summarized in Table 4.1. The results suggest a value of about 1.1×10^{16} g Si y^{-1}. Heath (1974) using a lower value for the annual global rate of production, adopting a Si:C ratio of 0.46 (by atoms) and allocating an arbitrary proportion of 50 and 70 % of the global photosynthesis to diatoms, calculated a range of values of from 7.9×10^{15} to 1.5×10^{16} g Si y^{-1}. The value calculated here is within the range but the degree of correspondence between the two estimates may indicate no more than different but compensating errors in the assumptions used. The data in Table 4.1 illustrate the relative magnitude of the production of biogenic silica in sub-polar regions, in Antarctica and in other areas of upwelling and fertile neretic waters which is compatible with the global distribution of amorphous silica in sediments shown in Fig. 4.4.

TABLE 4.1 Summary of estimates of annual production of biogenous silica by diatoms.

Area Type	Production of algal carbon (μg at C y^{-1})	C:Si ratio (by atoms)	Production of biogenous silica (μg at Si y^{-1})	(g Si y^{-1})
Oligotrophic	3.3×10^{13}	0.015	5.0×10^{11}	1.4×10^{13}
Transitional	1.7×10^{14}	0.20	3.4×10^{14}	9.5×10^{14}
Oceanic divergencies and sub-polar Ocean	4.8×10^{14}	0.30	1.4×10^{14}	3.9×10^{15}
Shelf areas including Antarctic and other areas of upwelling	5.2×10^{14}	0.45	2.3×10^{14}	6.4×10^{15}

The formation of amorphous silica by radiolaria is dependent on the supply of organic matter from primary production at the surface. Only a small fraction of the organic matter produced by the phytoplankton is likely to be available for radiolaria and the ratio between the amorphous silica in these organisms and carbon produced by photosynthesis must be much lower than that appropriate to diatoms. Radiolaria are most abundant in those areas of the ocean which at the most contribute about only 50% of global photosynthesis and these organisms can presumably therefore only contribute a small proportion to annual global production of amorphous silica by diatoms.

4.5.2   The Dissolution of Biogenous Silica in the Oceans

4.5.2.1   Total addition of silicic acid to deep water

The concentrations of silicic acid in the oceans (Section 4.2.1.2) imply that dissolution of amorphous silica will be possible even in the deep water of the northern Pacific Ocean where the concentrations of silicic acid reach their maximum values. In this respect amorphous silica differs from calcium carbonate, particles of which may be prevented from dissolution in the upper parts of the water column by the saturation of the water with respect to calcium carbonate. However, as discussed in Section 4.4.2, various secondary processes may decrease the solution rates of detrital amorphous silica.

A proportion of the dissolution of amorphous silica occurs concurrently with the biological oxidation of the organic matter in the cells of diatoms (Grill and Richards, 1964). Part of this dissolution will occur in the immediate surface layers and will be available for reassimilation. A proportion of the detrital material will sediment to depths where reassimilation by diatoms is prevented by insufficient light and this fraction will not be available for recycling unless advective movement of these deeper layers redistributes silicic acid in solution to the well-illuminated surface layers and in many cases, therefore, the silica can remain unavailable for utilization by diatoms for long periods.   Appreciable removal of silicic acid from solution in deep water by radiolaria may be expected in some areas, although this cannot be observed directly from vertical profiles of the concentrations of silicic acid. Grill (1970) has described a model of the dissolution of particulate biogenous silica during sedimentation and the vertical flux of particulate carbon which provides the food source of radiolaria. The expression describing the vertical distribution of silicic acid can be fitted to observed vertical profiles of silicic acid by suitably adjusting the values of the derived model constants used in the expression. This then allows calculation of the ratio of the difference between silicic dissolution and silicic acid utilization rates to the coefficient of eddy diffusivity at different depths. Although the resultant of the dissolution and utilization rate cannot be calculated, the sign of the ratio predicted by the model signifies net uptake or removal. In some water columns the ratio is positive at all depths, but in others the ratio is negative at some depths suggesting formation of biogenic silica by radiolaria. The prediction of the model may be susceptible to confirmation by application of the stable silicon isotope techniques used by Goering *et al.* (1973), and such measurements might provide quantitative data of the rates of formation of biogenic silica by radiolaria.

The dissolution of biogenic silica which accompanies biological oxidation of the cellular organic matter of diatoms will ultimately be accompanied by the regeneration of phosphorus and inorganic nitrogen and an equivalent utilization of oxygen. The instantaneous rates of these different processes may be different because, for example, there is a tendency for phosphorus to be regenerated more rapidly than nitrogen. Over a sufficiently long period however the total regeneration of these elements will be expected to be related by constant ratios to the oxygen consumption, and the values of the ratios $\Delta O : \Delta Si : \Delta N : \Delta P$ of $106:16:16:1$ may apply in many areas (see Section 4.3.1).

A Si:P ratio of about 16:1 in sub-surface water containing less than about 1000 to 1500 µg Si l^{-1} is implied in the data presented by Berger (1970), but in water with higher concentrations of silicic acid these are disproportionately small or with no further increases in the concentration of orthophosphate in solution (Fig. 4.8). These patterns imply continuing enrichment with silicic acid after all the organic matter has been oxidized. There are considerable differences in the shapes of silicon-phosphate plots for samples taken in different oceans (Fig. 4.9). The shapes of these plots reflect the differences in the recruitment and dissolution of particulate silicon to the deep water in the different oceans (see Section 4.2.1.2), the extent of vertical mixing and advective upwelling in bringing deep water containing high concentrations of silicic acid into the upper layers and the relative importance of concurrent dissolution of biogenous silica during the oxidation of organic matter compared with post oxidative processes.

Berger (1970) made estimates of the rates of enrichment of deep water with silicic acid during the post oxidative stage and suggests a total global annual rate of about 2.0×10^{15} g Si y^{-1}. The data from the Atlantic which Berger used may underestimate the total enrichment of deep Atlantic water with silicate (Mann *et al.*, 1973, Fig. 2) and the contributions of silicic acid during the post oxidation phase may therefore on this basis be about 2.5 times greater than the value used by Berger.

The silicon-phosphorus plots for the North Pacific are more difficult to interpret. The slope of the plots over the range where silicic acid concentrations are phosphate related superficially suggest a Si:P ratio (by atoms) of at least 45. This seems very high if it does reflect the elementary ratios in the original cellular organic matter. Presumably the slope of these plots is the result of considerable mixing of water containing higher concentrations of silicic acid with other water in which the effects of concurrent dissolution of silicic acid and oxidation of organic matter are significant. Taking the plot of data for the central Pacific it might be reasonable to use a concentration of about 50 mg at M^{-3} as the extent of dissolution during the oxidative phase. This would suggest that the contribution of other processes may be about 1.5 times greater than Berger's estimate. The total contribution of silicic acid to the deep water in the post oxidation phase of dissolution calculated in this way might therefore be raised to over 3×10^{15} g Si y^{-1}. The calculation of the rates of enrichment of deep water by this method also depends on using a value for the residence time of the deep water. Berger used residence times based on ^{14}C dating of particulate carbonate and this may also produce some uncertainties in the estimate.

Other approaches to the estimation of the rates of dissolution of silica in deep water have been used by Kido and Nishimura (1973, 1975). Assuming that the Japan Sea could be considered as a closed system these workers

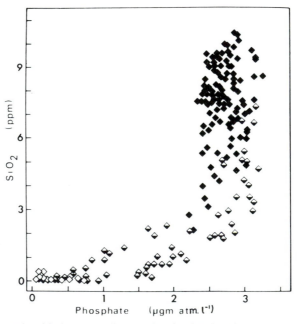

FIG. 4.8   The relationship between silicic acid and orthophosphate in water samples from the central Pacific Ocean. ◇ samples from upper 100 m. ◈ samples from 100–1000 m. ◆ samples from 1000 m (from Heath, 1974 after Berger, 1970).

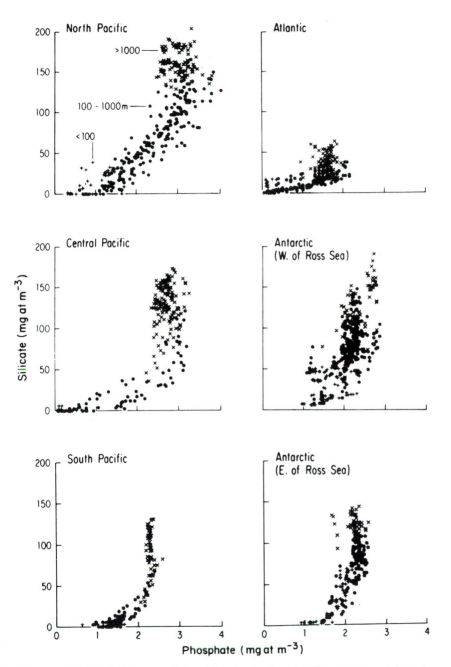

FIG. 4.9 Relationship between silicic acid and orthophosphate in various oceanic regions (from Berger, 1970).

estimated a rate of 2.0×10^{-6} g Si l^{-1} y^{-1} for non-oxidative dissolution in this area. In a later paper (Kido and Nishimura, 1975) they described a model based in the vertical distribution of particulate silica in the same area which gave estimates for the rate of dissolution in the water between 50 and 400 m in the range 1.1×10^{-6} and 1.1×10^{-5} depending on the values adopted for the sedimentation rates and vertical eddy diffusion. A similar value (Kido and Nishimura, 1973) of 1.6×10^{-6} g Si l^{-1} y^{-1} has been calculated for non-oxidative dissolution in the deep water of the Pacific Ocean. This is comparable with the value of 2.8×10^{-6} g Si l^{-1} y^{-1} which can be calculated from Berger's estimates of the rates of enrichment of the deep water in the Pacific Ocean.

4.5.2.2   Sources of silicic acid in deep water

The estimate of the rate of enrichment of deep water with silicic acid discussed in Section 4.5.2.1 cannot distinguish between several processes which may add silicic acid to this water, and Heath (1974) has considered a number of such possible sources. His estimates for the annual rates of deuteric and hydrothermal sources and for the halmrolysis of basaltic material are 2.5×10^{12} and 3.8×10^{13} g Si y^{-1} respectively. This represents only 1.5–3.0 % of the estimates in Section 4.5.2.1 of the total enrichment of deep water with silicic acid (see also Chapter 2, Section 2.2).

There has been a large number of analyses of the silicic acid concentrations in the interstitial water in deep-sea sediments (Fanning and Pilson, 1971) but, although there are some problems in using the data for comparative purposes because of variation in the methodology used in removing the interstitial water from the sediments, there are undoubtedly often considerably higher concentrations of silicic acid in interstitial water than in the overlying bottom water. Schink et al. (1974) have described the geographic variation in the interstitial silicic acid concentrations in sediments and these correlate well with geographical variations in the annual rates of production of biogenous silica (see Sections 2.2 and 5.1). Interstitial concentrations of silicic acid can be up to 15 000 µg Si l^{-1} greater than in the overlying water but the solubility of silicic acid at in situ temperatures and pressure (Figs 4.6 and 4.7) is considerably greater than this. On the other hand biogenous silica is present at all depths in such sediments and the increased concentrations of silicic acid in the interstitial water must, in part, be the result of dissolution of detrital biogenous silica which has been buried. This process and the diffusion of silicic acid from the sediments into the overlying water establish vertical gradients of concentration of silicic acid in the sediments, these concentrations reaching asymptotic values at some distance below the sediment surface.

Hurd (1973) fitted observed gradients of silicic acid in the interstitial water in sediments to a model which included the rate of solution of biogenous silica (see Section 4.4.2) and the apparent diffusion constant of silicic acid in the sediment. Fitting the expression to observed concentration gradients yielded an apparent diffusion coefficient of $2.4 \times 10^{-6}\,\text{cm}^{-2}\,\text{s}^{-1}$. Fanning and Pilson (1974) made experimental determinations of the flux of silicic acid from sediment cores and determined an effective diffusion coefficient of $3.3 \times 10^{-6}\,\text{cm}^{-2}\,\text{s}^{-1}$.

Schnink et al. (1975) have also described a model of the vertical variation of silicic acid in sediment which includes a term to describe the effects of mechanical agitation of the sediments caused by the burrowing activity of benthic animals. The model predicts that the asymptotic depth of silicic acid concentration is directly related to the rates of input of biogenous silica to the surface of the sediment. The asymptotic depth is also sensitive to mechanical disturbance of the sediments, increases in this causing steeper gradients and a decrease in the asymptotic depth.

The uniform concentrations of silicic acid in sediments below a particular depth in sediments thus seem to be controlled, at least in part, by diffusion and other physical factors rather than by the solubility of the biogenous silica in these regions of the sediment. However, effects on the rate of solution as a result of changes in crystallinity or secondary chemical processes such as the adsorption of other minerals cannot be completely excluded. It is also possible that other solid phases make some contribution to the pool of silicic acid in solution but this seems likely to be small.

The recorded values for the effective diffusion coefficient of silicic acid in sediments are similar to the value used by Heath (1974) to estimate the flux of silicic acid from sediments into the overlying water. The main uncertainties in making an estimate of the flux are the differences in the concentration gradients between the two reservoirs and there is insufficient data to allow estimation of a realistic global average weighted to allow for geographical differences in the interstitial concentrations of silicic acid. Heath adopted a value of $4.6 \times 10^{-3}\,\text{g Si l}^{-1}$ as a mean difference in silicic acid concentration which seems reasonable in the light of the values of between 2.8×10^{-3} to 2.8×10^{-2} reported by Fanning and Pilson (1971). Using the former value, Heath estimated a global flux of silicic acid from the sediments of $2.6 \times 10^{14}\,\text{g Si y}^{-1}$.

The estimates of other contributions of silicic acid to deep water suggest that the total contribution not associated with the oxidative phase of regeneration is between about 1.5 to $3.5 \times 10^{14}\,\text{g Si y}^{-1}$. There must be uncertainty about the proportion of the dissolution which occurs during the transit of the particles through the water column and how much occurs on the surface of the sediment before the particles are buried. Sedimentation rates of diatoms range from less than 1 for some types to 15 m day^{-1} for robust species like *Coscinodiscus* and cover a range of from 50 to 500 m day^{-1} for radiolaria (Berger, 1976). The effective sedimentation rates of silicous skeletons may be modified considerably if they are incorporated in larger particles of organic detritus. Lal and Lerman (1973) have pointed out that sinking rates and solution rates will be expected to decrease as a result of the reduction in size and surface area of particles. It is also possible for the sedimentation processes to be reversed in regions of intense advective upwelling.

Heath (1974) estimated the proportion of the total dissolution of bio-genous silica which occurs during the concurrent oxidation of organic matter by the difference between the total and non-oxidative dissolution. The value he suggested of 9.2×10^{14} g Si y^{-1} represents 82% of the total dissolution. This would imply a Si:C ratio (by atoms) of 0.28 in the original cellular material which seems high because it must relate to the total phytoplankton carbon. Calculation of the proportion of dissolution during the oxidation phase of regeneration in this way depends greatly on the value adopted for the total enrichment of deep water by non-oxidative dissolution. It has been suggested in Section 4.5.2.1 that Berger's estimate of this may be low. Using the values suggested in this chapter of 3.0×10^{15} g Si y^{-1} for the total enrichment of deep water with silicic acid, a value of 2.8×10^{14} g Si y^{-1} is suggested for the contribution of the dissolution during the oxidative phase. This represents 73% of the total dissolution of biogenic silica. If 3.0×10^{16} is used as the carbon content of the original cellular material (see Section 5.1), this implies a C:S ratio (by atoms) of 0.16 which seems more realistic (Section 4.3.1; see also Chapter 2, Sections 2.2 and 2.3).

4.5.3   Global fluxes of biogenous silica in the oceans

The estimated values of the annual flux of biogenous silica in the oceans are summarized in Fig. 4.10. The values used for the geochemical fluxes and input from terrestrial sources are those suggested by Heath (1974) and the biogeochemical fluxes are those suggested in the earlier section of this chapter. These values are not presented as alternatives which necessarily have greater relationship to reality than those suggested by Heath but are more consistent with other aspects of the biochemistry of silicon in the oceans which were not a part of Heath's model (see also Chapter 2).

The magnitude of the biogeochemical processes which remove and regenerate silicic acid in the oceans clearly dominate the other inputs and geochemical processes. This is reflected in the distribution of silicic acid in solution in the oceans, biogeochemical activity maintaining low concentrations in the surface layers. These, in conjunction with the similar behaviour and distribution of combined inorganic nitrogen and phosphorus, are probably the dominant factors in controlling the fertility of surface waters.

The site of dissolution is of great importance to the fertility of the surface layers and the various models adopt different boundaries to define deep water. Wollast (1975) suggests that about half of the silicic acid that is returned to the surface layers is carried by eddy diffusion. In this case, a large proportion of the dissolution must occur in the first several hundred metres of the water column.

Additionally the distribution of silicic acid must have an important role in the biogeographical distribution of diatoms. In contrast, the ecology of radiolaria may be controlled more importantly by physical factors such as temperature. The production of appreciable quantities of biogenous silicon by these protozoa in the areas of the ocean in which the surface waters are relatively poor in silicic acid is probably the cause of the fragile skeletons

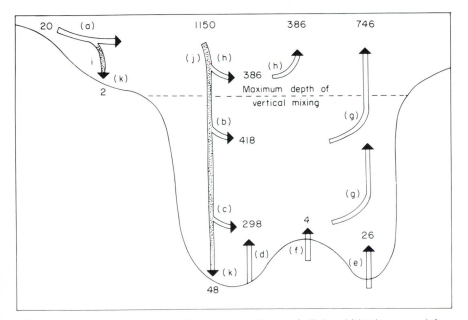

FIG. 4.10   Fluxes of biogenous silica and other fluxes of silicic acid in the ocean (after Heath (1974) and Wollast (1974) modified as suggested in this chapter). Fluxes of silicic acid in solution: (a) River input; (b) Oxidative phase dissolution; (c) non-oxidative phase dissolution; (d) dissolution from surface of sediment; (e) diffusion from sediment; (f) deuteric, hydrothermal and basalt dissolution; (g) upwelling; (h) eddy diffusion. Particulate fluxes: (i) sorption; (j) biological production of biogenous silica; (k) deposition. Magnitude of fluxes in $g \times 10^{13} \, y^{-1}$.

which characterize species found in the surface layers. In contrast, the species inhabiting deep water where higher concentrations of silicic acid are found have more robust skeletons. It is interesting that if the low concentrations of silicic acid in solution in tropical surface water tend to discriminate against diatoms that radiolaria are apparently able to acquire adequate supplies of silicon. The radiolaria are particle feeders in contrast to the strictly osmotrophic diatoms and may not be dependent on mobilizing silicic acid in solution but are able to ingest particulate silica.

There can be no doubt of the overwhelming importance of biological activity and the subsequent dissolution of the biogenous silica in controlling the concentrations and distribution of silicic acid in the oceans. There is conflicting evidence regarding the addition or removal of silicon in terrestrial drainage and some other geochemical processes such as the solution of glacial rock floor (Schutz and Turekian, 1965), but in any case the magnitude of such processes seems likely to be very small in relation to the biogeochemical fluxes. The same considerations seem likely to apply to the addition of silicic acid by deuteric, hydrothermal, and halmyrolytic processes.

The relative importance of the dissolution of biogenous silica during the oxidation and post-oxidation phase is difficult to assess with any certainty. It is equally difficult to assess how much of the non-oxidative dissolution of particulate amorphous silica occurs on the surface of the sediments, but this must vary considerably in different areas and be related to some extent to the depth of the water column. Equally, it seems likely that the sinking rate of many of the skeletons of radiolaria will be considerably greater than that of diatoms and a greater proportion of such siliceous detritus may dissolve on the surface of the sediments than is the case with diatoms from the same environment. The possibility that the sinking rate of diatoms may be increased by inclusion in larger particles of organic detritus such as faecal pellets has been suggested as a factor which may accelerate sedimentation. However as noted in Section 4.4.2 such a micro-environment may accelerate dissolution and in any case the influence of inclusion in organic detrital particles cannot extend beyond the area where dissolution is associated with oxidative processes.

One of the most striking more recent contribution to our knowledge of the biogeochemical cycling of silicon in sea water is perhaps the realization of the magnitude of the flux of silicic acid from sediments. This is of the same order as the contribution to the oceans of silicic acid by terrestrial drainage. Although there are considerable difficulties in allocating global values to this flux because of a lack of suitable data of the geographical valuations in the concentration of silicic acid in the pore water of sediments, it seems unlikely that more refined data will have a great influence on present estimates of the importance of this flux.

REFERENCES

Alexander, G. B., Heston, W. M., and Iler, R. K. (1954). *J. phys. Chem.*, **58**, 453.

Antia, N. J., McAllister, C. D., Parsons, T. R., Stephens, K., and Strickland, J. D. H. (1963). *Limnol. Oceanogr.*, **8**, 166.

Armstrong, F. A. J. (1958). *J. mar. biol. Ass. U.K.*, **17**, 23.

Armstrong, F. A. J. (1965). *In* "Chemical Oceanography", Vol. 1 (J. P. Riley and G. Skirrow, eds.). Academic Press, London and New York.

Berger, W. H. (1968). *Science*, **159**, 1237.

Berger, W. H. (1970). *Bull. geol. Soc. Amer.*, **81**, 1385.

Berger, W. H. (1976). *In* "Chemical Oceanography", 2nd Ed., Vol. 5 (J. P. Riley and R. Chester, eds.). Academic Press, London and New York.

Broecker, W. S. (1974). "Chemical Oceanography". Harcourt Brace Jovanovich, Inc., New York.

Bruyevich, S. and Ivanenkov, V. N. (1971). *Oceanology*, **11**, 699.

Chisholm, S. W., Azam, F., and Eppley, R. W. (1978). *Limnol. Oceanogr.* **23**, 518.

Clowes, A. J. (1938). *"Discovery" Rep.*, **19**, 1.

Conway, H. L. and Harrison, P. J. (1977). *Mar. Biol.*, **34**, 33.

Copin-Montegut, C. and Copin-Montegut, G. (1972). *Deep-Sea Res.*, **19**, 445.

Copin-Montegut, C. and Copin-Montegut, G. (1978). *Deep-Sea Res.*, **25**, 911.

Darley, W. M. and Volcani, B. E. (1969). *Exptl. cell. Res.*, **58**, 334.

Davies, C. O. (1976). *J. Phycol.*, **12**, 291.

Droop, M. R. (1973). *Amer. Zool.*, **13**, 209.

Durbin, E. G. (1977). *J. Phycol.*, **13**, 150.

Eppley, R. W., Jolmes, R. W., and Paasche, E. (1967). *Arch. Microbiol.*, **56**, 305.

Fanning, K. A. and Pilson, M. E. Q. (1971). *Science*, **173**, 1228.

Fanning, K. A. and Pilson, M. E. Q. (1974). *J. geophys. Res.*, **79**, 1293.

Goering, J. J., Nelson, D. M., and Carter, J. A. (1973). *Deep-Sea Res.*, **20**, 777.

Goldberg, E. D., Baker, M., and Fox, D. L. (1952). *J. Mar. Res.*, **11**, 194.

Greenberg, S. A. (1957). *J. phys. Chem.*, **61**, 960.

Greenberg, S. A. and Price, E. W. (1957). *J. phys. Chem.*, **61**, 1539.

Gregg, J. M., Goldstein, S. T., and Walters, L. J. (1977). *J. Sed. Petrol.*, **47**, 1623.

Grill, E. V. (1970). *Deep-Sea Res.*, **17**, 245.

Grill, E. V. and Richards, R. A. (1964). *J. mar. Res.*, **22**, 51.

Guillard, R. R. L., Kilham, P., and Jackson, T. A. (1973). *J. Phycol.*, **9**, 233.

Harrison, J. P., Conway, H. L., and Dugdale, R. C. (1976). *Mar. Biol.*, **35**, 177.

Harrison, P. J., Conway, H. L., and Holmes, R. W. (1977). *Mar. Biol.*, **43**, 19.

Heath, G. R. (1974). *In* "Studies in Paleo-Oceanography" (Hay, W. W., ed.). Soc. Econom. Paleont. Mineral. Sp. Pub. 20.

Heckey, R. E., Mopper, K., Kilham, P., and Degens, E. T. (1973). *Mar. Biol.*, **19**, 323.

Henderson, M. E. K. and Duff, R. B. (1963). *J. Soil. Sci.*, **17**, 236.

Hurd, D. C. (1972). *Earth Planet. Sci. Letters*, **15**, 411.

Hurd, D. C. (1973). *Geochim. Cosmochim. Acta*, **37**, 2257.

Hurd, D. C. and Theyer, F. (1975). *In* "Analytical Methods in Oceanography" (Gibb, T. R. P., ed.). Ad. Chem. Series No. 147. Amer. Chem. Soc. Washington, D.C.

Jones, M. M. and Pytkowicz, R. M. (1973). *Bull. Soc. R. Sci. Liege*, **42**, 118.

Jørgensen, E. G. (1955). *Physiol. Plant.*, **8**, 846.

Kamatani, A. (1969). *J. oceanogr. Soc. Japan*, **25**, 63.

Kamatani, A. (1971). *Mar. Biol.*, **8**, 89.

Kato, K. and Kitano, Y. (1968). *J. oceanogr. Soc. Japan*, **24**, 147.

140   *C. P. Spencer*

Kido, K. (1974). *Mar. Chem.*, **2**, 263.
Kido, K. and Nishimura, M. (1973). *J. oceanogr. Soc. Japan*, **29**, 185.
Kido, K. and Nishimura, M. (1975). *Deep-Sea Res.*, **22**, 323.
Koblentz-Mishke, O. J., Volkovinski, V. V., and Kabanova, J. G. (1970). In "Scientific Exploration of the South Pacific" (Wooster, W. S., ed.). Nat. Acad. Sci. Washington, DC.
Krauskopf, K. B. (1956). *Geochim. Cosmochim. Acta*, **10**, 1.
Krauskopf, K. B. (1959). *Soc. Econ. Paleont. Mineral. Sp. Publ.*, **7**, 4.
Lal, D. and Lerman, A. (1973). *J. geophys. Res.*, **78**, 7100.
Land, L. S. (1976). *J. Sed. Petrol.*, **46**, 967.
Langerstrom, G. (1959). *Acta Chem. Scand.*, **13**, 722.
Laumers, A. M. and Heinen, W. (1974). *Arch. Microbiol.*, **95**, 67.
Lewin, J. C. (1961). *Geochim. Cosmochim. Acta*, **21**, 182.
Lewin, J. C. (1965). *Naturwissenschaften*, **52**, 70.
Lewin, J. C. (1966). *J. Phycol.*, **2**, 160.
Lisitzin, A. P. (1967). *Int. geol. Rev.*, **9**, 631.
Lisitzin, A. P. (1970). Nat. Acad. Sci. Washington, DC.
Lisitzin, A. P. (1972). Soc. Econ. Paleont. Mineral. (Rodolfo, K. S., ed.). Sp. Publ. 17.
Lisitzin, A. P., Belyaev, Y. I., Bogdanov, Y. A., and Bogoyavlenskiy, A. N. (1966). Oceanological Research. (Results of I.G.Y. Section X), **10**, 1.
Livingstone, D. A. (1963). *U.S. Geol. Survey Prof. Paper*, **440**.
McAllister, C. D., Parsons, T. R., Stephens, K., and Strickland, J. D. H. (1961). *Limnol. Oceanogr.*, **6**, 237.
Mann, C. R., Coote, A. R., and Garner, D. H. (1973). *Deep-Sea Res.*, **20**, 791.
Mehard, C. W., Sullivan, C. W., Azam, F., and Volcani, B. E. (1974). *Physiol. Plant*, **30**, 265.
Menzel, D. W. and Ryther, J. H. (1960). *Deep-Sea Res.*, **6**, 351.
Nelson, D. M. and Goering, J. J. (1977). *Deep-Sea Res.*, **24**, 65.
Nelson, D. M., Goering, J. J., Kilham, S. S., and Guilliard, R. R. L. (1976). *J. Phycol.*, **12**, 246.
O'Connor, T. L. and Greenberg, S. A. (1958). *J. Phys. Chem.*, **62**, 1195.
Okamato, G., Okura, T., and Goto, K. (1957). *Geochim. Cosmochim. Acta*, **12**, 123.
Paasche, E. (1973a). *Mar. Biol.*, **19**, 262.
Paasche, E. (1973b). *Mar. Biol.*, **19**, 117.
Park, K. (1967). *Limnol. Oceanogr.*, **12**, 353.
Parsons, T. R., Stephens, K., and Strickland, J. D. H. (1961). *J. Fish. Res. Bd. Can.*, **18**, 1001.
Platt, T. and Subba Rao, D. V. (1973). Tech. Rep. No. 370. Fish. Res. Bd. Can.
Purushothamam, A., Chandramohan, D., and Natarajan, R. (1974). *Current Sci.*, **43**, 282.
Richards, F. A. (1958). *J. Mar. Res.*, **17**, 449.
Round, F. E. (1973). "The Biology of the Algae". Edward Arnold, London.
Schink, D. R., Fanning, K. A., and Pilson, M. E. Q. (1974). *J. geophys. Res.*, **79**, 2243.
Schink, D. R., Guianasso, N. L., and Fanning, K. A. (1975). *J. geophys. Res.*, **80**, 3013.
Schmid, A. M. M. and Schulz, D. (1979). *Protoplasma*, **100**, 267.
Schott, F. and Ehrhardt, M. (1969). *Kieler Meeresforch*, **25**, 272.
Schutz, D. F. and Turekian, K. K. (1965). *Geochim. Cosmochim Acta*, **29**, 259.
Siever, R. (1962). *J. Geol.* **70**, 127.
Stefansson, U. (1968). *Deep-Sea Res..* **15**, 541.

Strickland, J. D. H. (1965). *In* "Chemical Oceanography", Vol. 1 (Riley, J. P. and Skirrow, G., eds.). Academic Press, London and New York.

Strickland, J. D. H., Holm-Hansen, O., Eppley, K. W., and Linn, K. J. (1967). *Limnol. Oceanogr.*, **14**, 23.

Sverdrup, H. N. (1953). *J. Cons. int. Explor. Mer.*, **18**, 287.

Thomas, W. H. (1975). *Deep-Sea Res.*, **22**, 671.

Toyoto, T. and Okabe, S. (1972). *J. oceanogr. Soc. Japan*, **23**, 1.

Tsunogai, S., Kusakabe, M., Iizumi, H., Koike, I., and Hattori, A. (1979). *Deep-Sea Res.*, **26**, 641.

Vinogradov, A. P. (1953). Translation by Sears Foundation for Marine Research. Yale University Press, New Haven, Connecticut.

Werner, D. (1977). *In* "The Biology of Diatoms" (Werner, D., ed.). Blackwell Scientific Publications, Oxford.

Werner, D. (1978). *In* "Drugs and Food from the Sea. Myth or Reality" (Kaul, P. N. and Sindermann, J., eds.). University of Oklahoma.

Willey, J. D. (1974). *Mar. Chem.*, **2**, 239.

Wollast, R. (1974). *In* "The Sea", Vol. 5 (Goldberg, E. D., ed.). Wiley-Interscience, New York.

Sedimentary Geochemistry of Silicon

S. E. Calvert

Department of Oceanography
University of British Columbia
Vancouver, British Columbia
Canada

5.1   INTRODUCTION

The central role played by silicon in earth's surface biogeochemical processes is demonstrated by the importance of sedimentary cycling of the element over geological time (Garrels and Mackenzie, 1971; see also Chapter 2), the crucial importance of silicic acid as a nutrient in the biological economy of the modern ocean (Dugdale, 1972; see also Chapter 4) and the accumulation of vast quantities of silica in the form of chert in the geological record (Pettijohn, 1957; Gruner, 1965). In these various reservoirs, silicon is involved in a number of chemical reactions, including the precipitation and solution of amorphous silica, the precipitation of crystalline forms of silica and their eventual recrystallization, and the formation of a wide variety of aluminosilicate minerals. The biological, chemical, and geological cycles of silicon interact in a complex way and the complexity of the reactions and reaction products has led to the production of a voluminous literature.

There has been a resurgence of interest in siliceous sedimentary deposits and in the marine geochemistry of silicon over the past decade. The new information, which is providing a more complete understanding of the processes which bring about the deposition and preservation of silicon in the geological record, comes from re-evaluations of the geochemical balance over geological time, from investigations of modern marine sedimentation and from discoveries made by the Deep Sea Drilling Project (JOIDES, 1967). The

143

emphasis in this chapter will be directed towards the marine geochemistry of silicon, with discussion of processes in lakes and of some recent work on Tertiary siliceous rocks on land.

5.2 DEFINITIONS

Silicon occurs in natural waters as the undissociated monomeric acid $Si(OH)_4$ (Bruevich, 1953; Sillén, 1961; see also Chapter 3). In much of the oceanographic literature it is referred to as silicate or reactive silicate. It will be referred to as *dissolved silicon* in this chapter. In solid phases of concern here, silicon occurs as a number of polymorphs of SiO_2, ranging widely in crystallinity; they will be referred to as *silica*.

The stable polymorph of SiO_2 at earth's-surface conditions is α-quartz (stable up to 573°C) and this is the most abundant mineral in sedimentary rocks. Its dominance here is due to its hardness (number 7 on the Mohs scale), lack of cleavage, and low solubility in water. It is derived from the weathering of quartzose igneous, metamorphic, and sedimentary rocks (see Chapter 2). Other polymorphs are stable at higher temperatures and include β-quartz (573–870°C), tridymite (870°–1470°C), and cristobalite (1470°–1710°C). Coesite (Coes, 1953; Sosman, 1954) and stishovite (Stishov and Popova, 1961) are high-pressure polymorphs formed by transient shock pressure (Frondel, 1962).

A number of poorly crystalline polymorphs of silica are also stable at earth's-surface conditions. Jones and Segnit (1971) have provided a classification of such phases. *Opal-A* is nearly amorphous; it has extremely short-range order and an X-ray diffraction pattern resembling that of glass. It is the constituent of most precious opals, diatomites, and geyserites and is the substance precipitated by organisms (diatoms, silicoflagellates, chrysophytes, radiolarians, grasses, and sponges). *Opal-CT* is characterized by some of the X-ray diffraction reflections typical of low cristobalite with some of the strong lines of tridymite. It is the main constituent of "common" opals and is widely distributed in relatively young siliceous sedimentary rocks, including bentonites and claystones, where it has been referred to as cristobalite, cristobalite-tridymite and lussatite. *Opal-C* is a more ordered form, having intense X-ray reflections of α-cristobalite and some minor reflections of tridymite which cause some line broadening. The form is rare in sedimentary rocks, being found in only a few deposits associated with lava flows.

Chert is a common siliceous sedimentary rock consisting of microcrystalline quartz (Folk and Weaver, 1952; Smith, 1960). *Chalcedony* is a fibrous variety of quartz common in some cherts, which may contain some amorphous silica (Donnay, 1936; Folk and Weaver, 1952). Because of the fine grain-size and the possible presence of amorphous silica, chert appears to be somewhat more soluble in water than macrocrystalline quartz.

Opal-CT has received a great deal of attention over the last decade because of its discovery in several geologically young cherts and in many siliceous rocks recovered by drilling in the deep ocean. It evidently forms as an intermediate phase in the conversion of amorphous silica to chert (see Section 5.4). Following Calvert (1971a), rocks consisting of opal-CT are referred to as *porcelanites*, a term originally coined by Taliaferro (1934) and used extensively by Bramlette (1946).

5.3 MARINE GEOCHEMISTRY OF SILICON

5.3.1 Dissolved Silicon

In the modern ocean, the cycling of silicon is governed by the complex interaction between biological activity and physical circulation (see Chapter 4). Silicon in sea water is derived mainly from continental denudation and from sea-floor hydrothermal activity, and is extracted by planktonic and benthonic organisms for the construction of cell walls and tests. The distribution in ocean waters is by no means uniform (see Chapter 4). The concentration of dissolved silicon is everywhere less than about 230 µmol kg^{-1} (~ 14 ppm SiO_2); it is generally low in surface waters, where most of the biological extraction takes place, and it increases with water depth where the silicon is returned to solution by the dissolution of settling biological debris. Individual profiles of concentration with depth may show mid-depth maxima caused by such regeneration (Grill, 1970) and these may be further enhanced by deep minima brought about by the advection of abyssal waters having lower dissolved silicon levels (see Craig *et al.*, 1972). In addition, some profiles show distinct increases towards the abyssal sea-floor signifying an additional supply from siliceous bottom sediments (Edmond *et al.*, 1979a).

There are marked differences in the concentration of dissolved silicon between the major ocean basins, which are brought about by the abyssal circulation (Stommel and Arons, 1960; Berger, 1970). The Pacific gains dissolved silicon and other plant nutrients (Redfield *et al.*, 1963) from the Atlantic and Indian Oceans and returns nutrient-depleted surface water. The concentration difference is maintained in spite of higher rates of supply of silicon to the surface Atlantic by run-off. The distribution of silicon in the oceans is therefore controlled by a combination of physical, chemical, and biological processes, the latter two having rates which are higher than the mixing time of the oceans.

5.3.2 Biological Extraction

Krauskopf (1956) first suggested that the low concentration of dissolved silicon in sea water compared with river water is caused by the activities of marine organisms. Two planktonic groups are important in this process, the primary producing algae (diatoms, silicoflagellates and chrysophytes) and the protozoan radiolarians. Siliceous sponges are only locally important.

The mechanism of uptake and precipitation of silicon in planktonic organisms is known only for the diatoms which can be grown successfully in laboratory cultures. A great deal of physiological and biochemical work (Lewin, 1954, 1955a,b, 1957, 1962; Coombs et al., 1967a,b,c,d; Coombs and Volcani, 1968; Darley and Volcani, 1969; Healey et al., 1967) and detailed observations of the ultra-fine structure of newly-divided diatom cells (Reimann, 1964; Reimann et al., 1965, 1966; Stoermer and Pankratz, 1964; Stoermer et al., 1964, 1965) has shown that silicon is a fundamental constituent of the cell wall and, moreover, is an essential requirement for cell division (for further discussion see Chapter 4). Lack of silicon in the culture medium causes lower rates of synthesis of nucleic acids, proteins, carbohydrates, chlorophyll, and fucoxanthin and lower rates of evolution of O_2, fixation of CO_2 and uptake of nitrogen and phosphorus (Werner, 1977). During silicon starvation, DNA synthesis is blocked and mitosis is inhibited. Silicon, in the form of undissociated monomeric silicic acid, is the only known species that can be utilized, simple substituted acids (e.g. $H_3SiO_4CH_3$) being inactive. Furthermore, silicon cannot be replaced by any element of similar physical or chemical properties; germanium at concentrations of 1–10 mg l^{-1} is a specific inhibitor to diatom growth (Lewin, 1966).

Diatom cell walls vary widely in their degree of silicification (Einsele and Grim, 1938), SiO_2 contents ranging from less than 10% to around 50% of the dry weight (Vinogradov, 1953; Lewin et al., 1958). As well as a species-dependent variation in silicification, the amount of silicon in the cell varies with the composition of the growth medium and the rate of cell division (Jørgensen, 1955); in silicon-deficient media, extremely thin cells are produced (Booth and Harrison, 1979) whereas under conditions of limited nutrient availability, other than silicon, thickened cell walls are formed (Lewin, 1957).

The diatom cell walls are composed of hydrated SiO_2, here referred to as *opal*. Chemical analyses of this material are sparse; older data show highly variable contents of Na, K, Fe, and Al (Rogall, 1939; Desikachary, 1957). More recently, Martin and Knauer (1973), using scrupulously cleaned samples, showed that the acid-leached residues of siliceous plankton contain significant amounts of Ti and Al, elements which would normally be considered indicative of aluminosilicate contamination.

X-ray diffraction studies of diatom and radiolarian shells (Calvert, 1966; Mizutani, 1966; Komatani, 1971) suggest that the silica has a structure similar to glasses (Warren and Biscoe, 1938; Cartz, 1964) and to precious opals (Jones *et al.*, 1964). The structure is probably a three-dimensional random network of tetrahedral SiO_4 units, similar to that deduced for vitreous silica (Bell and Dean, 1972). There have been sporadic reports of crystalline phases in biogenous silica, including quartz, α-tridymite, and α-cristobalite (Brandenberger and Frey-Wyssling, 1947; Lanning *et al.*, 1958; Sterling, 1967) but these are in all probability produced either during sample preparation or by contamination (Jones and Milne, 1963).

Detailed studies of the ultra-fine structure and the physical properties of diatom and radiolarian shells, of a wide range of ages, have been reported by Hurd and Theyer (1975, 1977). Further information is given in Chapter 6.

Information presently available, and reviewed in Chapter 4, suggests that diatoms, the most important primary producers in the sea, are responsible for the bulk of the extraction of dissolved silicon from ocean waters. Primary production in surface waters varies regionally (Gessner, 1959), caused in turn by physical processes governing circulation, mixing, and stability. Around Antarctica, intense winter mixing and the effects of the circumpolar current combine to bring about vertical exchange of surface and deep water (Deacon, 1963) which results in the production of fertile surface waters throughout the year (Bogoyavlenskiy, 1967). Conditions are therefore adequate for sustained production whenever solar radiation is sufficient. In the sub-Arctic regions, winter mixing over much shallower areas produces the same effect; however, slow upwelling of deep Pacific water (Krauss, 1962), which is nutrient-rich (Edmond *et al.*, 1979a), augments the supply of nutrients to the surface layers.

At low latitudes along the eastern boundaries of the oceans, conditions are favourable for coastal upwelling due to the transport of surface water away from the coastline under the influence of the trade winds. The upwelled water is derived from a depth of a few hundred metres (Sverdrup, 1938), which in all cases is sufficient to cause the surface waters to be enriched in plant nutrients. The California, Peru, Canary, and Benguela Currents are well-known examples of highly productive boundary upwelling environments. Open ocean upwelling also occurs in the equatorial regions, especially in the Pacific, once again in response to wind-induced surface water divergence.

Certain coastal gulfs and embayments are also intermittently highly productive due to the process of upwelling. Thus, the Sea of Okhotsk (Bezrukov, 1955) and the Gulf of California (Calvert, 1966) in the Pacific are both sites of high diatom production.

Elsewhere in the oceans, the stability of the water column precludes

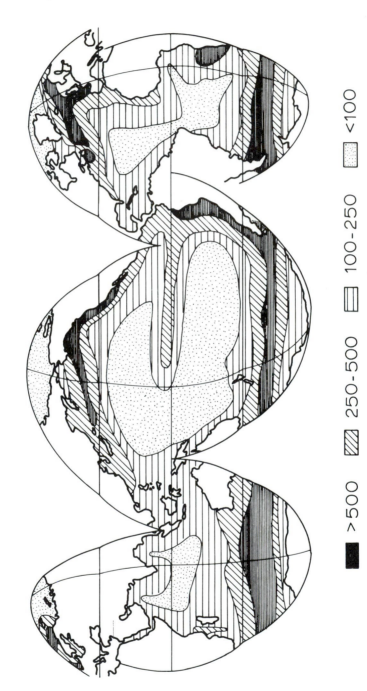

FIG. 5.1 Variation in the rate of extraction of dissolved silicon (g SiO$_2$ m^{-2} y^{-1}) by phytoplankton in near-surface waters of the ocean. Modified from Lisitzin et al. (1967).

■ >500   ▨ 250-500   ▥ 100-250   ▦ <100

vertical exchange. Consequently, primary production and standing crops are low, especially in the mid-latitude gyres in the major ocean basins.

Lisitzin *et al.*, (1964) have mapped the regional variation in the uptake of silicon by primary producers from the measured production rates and the SiO_2/C ratios in bulk plankton (Fig. 5.1). This is assumed to reflect the activities of diatoms; that this is a reasonable interpretation is shown by the recent work of Turpin and Harrison (1979) showing that diatoms have a competitive advantage over other phytoplankters where the specific nutrient supply rate is high, as in an upwelling environment.

The total rate of extraction of silicon by diatoms (used synonymously for siliceous phytoplankton), expressed as SiO_2, has been given by Heath (1974) as $1.7–3.2 \times 10^{16}$ g y^{-1}. An unknown additional, probably large, mass is used by radiolarians and sponges. This figure should be compared with a rate of supply of silicon by river run-off of 3.3×10^{14} g SiO_2 y^{-1} and an estimated flux of 1.9×10^{14} g SiO_2 y^{-1} from seafloor hydrothermal activity (Edmond *et al.*, 1979b) (see Section 5.4). The comparison shows that the total mass of silicon in the ocean is rapidly recycled in order to sustain the production. Furthermore, the residence time of dissolved silicon in the biological cycle is much shorter than the circulation time of the entire ocean of around 1000 years (Section 5.4).

5.3.3 Regeneration

It is well known that fragile diatom shells dissolve rapidly after growth ceases and the protective coating of the silica wall is lost (Lewin, 1961). This loss of biogenous silica is shown by the decrease in the concentration of diatom shells with depth (Gilbert and Allen, 1943; Round, 1968), the change in the composition of the flora with depth (Gilbert and Allen, 1943) and the decrease in the concentration of suspended amorphous silica with depth (Lisitzin *et al.*, 1967). The rapid solution of silica shells in the upper layers is evidently controlled by the oxidative regeneration of protoplasm in the water column, because the increase in dissolved silicon with depth is partly correlated with the increase in dissolved phosphate, first noticed by Richards (1958), supported by the modelling results of Grill (1970), and discussed by Berger (1970) and Kido and Nishimura (1973), itself a product of the oxidation of organic matter. Another fraction of the particulate silica dissolves deeper in the water column and on the sea-floor because here dissolved silicon increases with depth while phosphate remains more or less constant. Because of the relatively low rate of solution of biogenous opal in sea water, partly due to surface coatings (Lewin, 1961), it is thought likely that a significant amount of solution takes place on the sea-floor rather than in the water column (Heath, 1974). Corroborative evidence for this is provided by the close correspondence between microplankton assemblages in surface waters and in the underlying sediments (Murray and Renard, 1891; Kanaya and Koizumi, 1966) and the fact that fine-grained radioactive debris reaches abyssal depths very quickly (Gross, 1967). Schrader (1971) has shown that diatom shells are delivered rapidly to the deep-sea floor in faecal pellets, and such a mechanism is probably responsible for the major part of the particulate flux in the ocean in large particles (McCave, 1975).

An important subsidiary effect of the solution of thinly-silicified diatom shells in the water column and the survival of the more robust forms in the underlying sediments is the distinct difference in the diatom thanatocoenoses compared with the biocoenoses (Kolbe, 1955; Calvert, 1966). Similarly, radiolarian thanatocoenoses differ from the biocoenoses (Petrushevskaya, 1971; Renz, 1973). In spite of such changes, distinct biogeographic provinces of diatoms (Kanaya and Koizumi, 1966; Kozlova and Mukhina, 1967; N. Maynard, 1976) and radiolaria (Casey, 1971; Goll and Bjørklund, 1971, 1974; Moore, 1978; Johnson and Nigrini, 1980) are recognized on the sea-floor.

Discussion of the relative importance of water column and sea floor solution of siliceous tests in the overall geochemical balance is given in Section 5.4.

5.3.4   Sedimentation

Although only a relatively small fraction of the silicon fixed in the euphotic zone reaches the sediments in abyssal depths, amorphous silica nevertheless comprises a significant part of marine sediments and this has been recognized since the results of the Challenger Expedition became available (Murray and Renard, 1891). This component is difficult to measure quantitatively since crystalline silica and aluminosilicates are ubiquitous components of virtually all sediments. Riedel (1959) obtained visual estimates of the weight percentage of biogenous silica in Pacific sediments and established the main distribution pattern of diatoms and radiolaria in this area.

More recently, quantitative data have become available from the application of wet chemical (Aleksina, 1960; Hurd, 1973), X-ray diffraction (Goldberg, 1958; Calvert, 1966; Eisma and Van der Gaast, 1971) and infrared absorption spectrophotometric (Chester and Elderfield, 1968) techniques. Table 5.1 provides a modest intercomparison of these techniques using a range of modern diatomaceous sediments. Boström *et al.* (1973) estimated opal contents of Pacific sediments from bulk SiO_2 and Al_2O_3 ratios and Leinen (1977) has extended this approach with a more sophisticated technique using bulk sediment major element analyses. This has been successfully applied to a regional study of silica sedimentation through the

TABLE 5.1   Comparison of analyses of opaline silica in marine
sediments (wt. %).

Sample[a]	Chemical extraction[b]	X-ray diffraction[c]	Infra-red absorption[d]
BS26	4.2	10	8
L56	18.8	32	27
BS35	35.5	40	39
L117	33.1	52	55
BS29	49.2	70	69

a. BS samples from Sea of Okhotsk, L samples from Gulf of
California; b. data kindly supplied by P. L. Bezrukov (Institute of
Oceanology, Moscow); c. data obtained using method in Calvert
(1966); d. from Chester and Elderfield (1968).

Cenzoic in the Antarctic Ocean (Brewster, 1980). Eggiman *et al.* (1979) have
completed a detailed re-evaluation of the wet chemical technique which
shows promise.

The principal sources of information on the distribution of opal in marine
sediments are derived from the work of Soviet workers using an early
chemical extraction method. Figure 5.2 shows a schematic distribution for
the deep ocean and for several nearshore areas. The highest concentrations
are found in abyssal depths around Antarctica and this material is
overwhelmingly diatomaceous. In the North Pacific, diatomaceous clays also
occur in the Bering Sea (Bezrukov, 1960) and in abyssal depths south of the
Aleutian Islands. Equatorial Pacific sediments contain significant amounts of
radiolaria, but opal values rarely exceed 10 % by weight on a carbonate-free
basis. Likewise, some radiolarian oozes are known to occur in the equatorial
Indian Ocean (Lisitzin, 1967).

The distribution of amorphous silica in suspension in sea water also
reflects the regional variation in primary production and the distribution of
siliceous sediments on the sea floor. This fraction of the suspended
particulate standing crop in ocean waters has been measured by Lisitzin *et al.*
(1967) and Kido (1974). In a longitudinal profile through the Pacific, the
highest concentrations of biogenous silica particles are found around the
Antarctic continent and north of 40°N, consistent with the pattern of fertility
of the surface waters.

In addition to the distribution of opal in abyssal sediments, silica also
accumulates at high rates in several nearshore areas (Fig. 5.3). The silica is
entirely diatomaceous and values can reach 70 % by weight of the total
sediment. The areas shown are all sites of high primary production brought
about by upwelling.

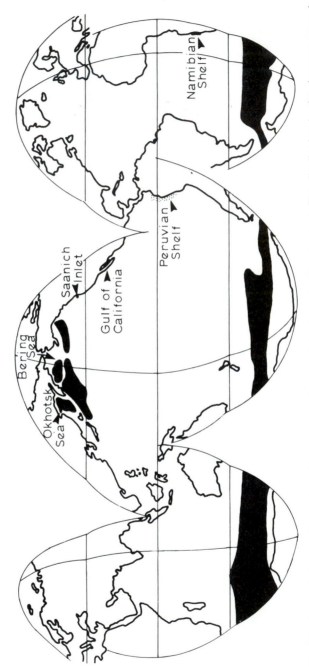

FIG. 5.2 Areas of the sea floor where biogenous opal accumulates to a significant extent. The distribution in the deep-sea areas is taken from Lisitzin (1967). Other marginal distributions are shown in Fig. 5.3.

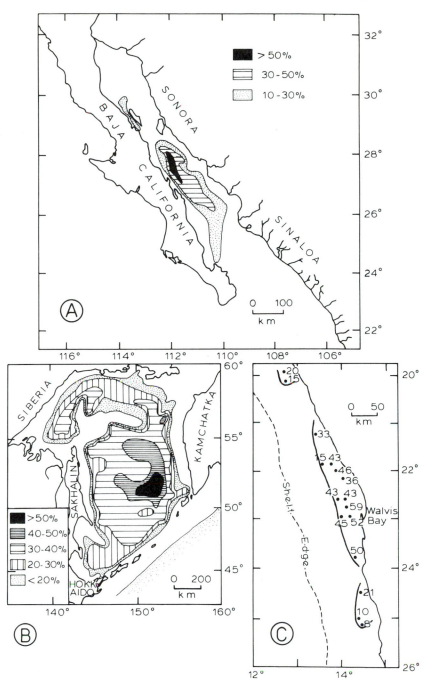

FIG. 5.3. Concentration of biogenous opal in surficial sediments of three marginal areas of the ocean. Data sources: (A) Gulf of California from Calvert (1966); (B) Sea of Okhotsk from Bezrukov (1955); (C) Namibian Shelf modified with revised data from Calvert and Price (1970). Values in % dry weight.

The distribution of biogenous opal in marine sediments (Figs 5.2 and 5.3) directly reflects the regional variation in primary production and dissolved silicon extraction by phytoplankton, as emphasized by Heath (1974). This in turn is controlled by purely physical processes of circulation, mixing, and the extent of vertical transport and stability in the water column. In the Gulf of California, for example, the high production is in response to upwelling brought about by seasonal changes in the wind patterns over the area (Calvert, 1966). The exchanges of water between the Gulf and the open Pacific result in the supply of silicon to the surface waters more than two orders of magnitude greater than that estimated to accumulate in the underlying sediments. In contrast to earlier notions that the masses of silica in sedimentary rocks could not be supplied from sea water, this evidence indicates that the oceanic reservoir is adequate and that well-known oceanographic processes concentrate silica in favourable nearshore environments which act as sinks for dissolved silicon from the open ocean.

5.4   THE MARINE BUDGET FOR SILICON

The importance of the biological cycle in controlling the abundance, distribution, and the budget of dissolved silicon in the sea has been emphasized by Riedel (1959), Harriss (1966), Gregor (1968), Calvert (1968), and Heath (1974). An extension of this proposal, whereby the proportions of the important nutrient elements in sea water are controlled by their relative abundances in the biochemical cycle, was presented by Redfield (1958). Alternatively, it is suggested, following the seminal paper by Sillén (1961), that inorganic reactions, involving dissolved silicon and particulate material, are the main control on the concentration of silicon in sea water (Mackenzie and Garrels, 1965, 1966; Mackenzie et al., 1967; Garrels and Mackenzie, 1971). This problem is discussed in Section 5.7.

A re-examination of the cycles and the budget of silicon in the ocean by Heath (1974), Wollast (1974), and DeMaster (1981) has reaffirmed the importance of biological activity in controlling the silicon level in the modern ocean. A summary budget, incorporating up-to-date information on supplies and removal mechanisms, is shown in Table 5.2. As in many other marine budgets, the dominant source of dissolved silicon to the oceans remains continental run-off; the figure derived from Livingston (1963) has recently been revised by analyses from a larger number of rivers by Meybeck (1979).

The second largest flux of silicon to the ocean is from submarine hydrothermal discharge. The figure given in Table 5.2 is based on the silicon content of hydrothermal spring waters of the Galapagos Rift extrapolated to the world-wide flow of sea water through the hot oceanic crust (Corliss et al.,

1979; Edmond *et al.*, 1979). This source has a much more secure foundation compared with the earlier estimates of supplies by submarine volcanism of 3×10^{10} g SiO_2 y^{-1} (Calvert, 1968). The figure of 8×10^{13} g SiO_2 y^{-1} from halmyrolysis (Heath, 1974) is obtained from estimates of the release of silicon from the low-temperature alteration of oceanic basalt.

The importance of glacial weathering on Antarctica in supplying very large quantities of silicon to the ocean is apparently not as great (8×10^{14} g SiO_2 y^{-1}) as that originally suggested by Schutz and Turekian (1965) and supported by Burton and Liss (1973). Edmond (1973) drew attention to the magnitude of the internal cycle in the Antarctic Circumpolar Current via biological activity and resolution which serves to obliterate the

TABLE 5.2   Silicon budget in the ocean.

		10^{14} g SiO_2 y^{-1}
1. *Supply*		
River influx		3.3[a]
Hydrothermal influx		1.9[b]
Sea floor weathering		0.8[c]
	Total	6.0
2. *Internal cycle*		
Biological extraction		100[d]–325[e]
Regeneration (water column and sea floor)		90[f]
Escape of pore waters		5[g]
3. *Removal*		
Deep sea		2.7[h]
Estuaries		0.66[i]
Continental margins		0.4[h]
	Total	3.76

a. using a river flow of 3.2×10^{19} g y^{-1} (Garrels and Mackenzie, 1971) and a mean silicon content of river water of 10.4 ppm (Meybeck, 1979); b. from Edmond *et al.* (1979); c. from Heath (1974); d. from Lerman and Lal (1977); e. using the world primary production estimates of Platt and Subba Rao (1975), including the Arctic and Antarctic Sea, an SiO_2/C ratio in bulk plankton of 2 and a diatom productivity of 60% of total primary production (see Heath, 1974); f. assuming 90% of the silica fixed is regenerated in the water column and on the sea floor, this accords with the order of difference of the values given by Heath (1974); g. from Fanning and Pilson (1974). This is close to a value of 5.7×10^{14} g y^{-1} given by Heath (1974); h. from DeMaster (1981)—most of the removal in the deep sea (2.2×10^{14} g y^{-1}) takes place around Antarctica; i. assuming 20% of the riverine flux (DeMaster, 1981) is removed (biologically or abiologically) in estuaries.

effects, if any, of additional local sources of silicon. Hurd (1977) has shown, moreover, that finely-ground rock particles either precipitate silica or dissolve very slowly in sea water containing an equivalent amount of dissolved silicon to that found in Antarctic water. Using this information, together with considerations of recent estimates of the mass of rock material actually eroded from the Antarctic continent, the size range of particles so produced, the processes of mixing around Antarctica which produce the relatively high silicon concentrations observed and the sedimentation rates around the continent required to deposit all the rock flour claimed to be supplied, Hurd concludes that glacially eroded rocks from Antarctica are not a significant source of dissolved silicon in the oceans.

Table 5.2 illustrates the scale of the biologically mediated subcycle of dissolved silicon in the overall budget. The role of biological activity can be shown by considering a simple model of uptake and removal of silicon to the deep ocean (Lal and Lerman, 1973; Lerman and Lal, 1977). In a surface layer (of the order 200 m thick) silicon is removed by biological activities and by mixing with deep water. In the deep water layer, silicon is removed by mixing with the surface layers. For such a two-layer system, the amounts leaving and being added to the two layers are equal:

$$(K_s + K_b)C_sV_s = K_dC_dV_d \tag{1}$$

where K_s, K_b, and K_d are the rate constants (y^{-1}) for removal from the surface layer by mixing, for biological uptake and for removal from deep water by mixing, respectively, C_s and C_d are silicon concentrations $(\mu mol\, cm^{-3})$ and V_s and V_d are thicknesses (cm) of the surface and deep layers, respectively. The biological removal of silicon from the surface layer, assuming $V_d = 3.6 \times 10^5$ cm, $C_d = 150\, \mu mol\, l^{-1}$, $C_s = 15\, \mu mol\, l^{-1}$ (average Pacific values) and $1/K_d$ (the residence time of deep water) = 1000 y, from

$$K_bC_sV_s = (C_d - C_s)K_dV_d \tag{2}$$

is $2.9\, mg\, SiO_2\, cm^{-2}\, y^{-1}$. Lerman and Lal (1977) also give a water column regeneration rate, from a model of the solution of a settling assemblage of biogenous particles of a range of grain sizes, as $1.6\, mg\, SiO_2\, cm^{-2}\, y^{-1}$. The remaining fraction which settles to the surface sediment is therefore $1.3\, mg\, SiO_2\, cm^{-2}\, y^{-1}$ which is the same figure as that given by Heath (1974). From these considerations, it appears that approximately 55% of the biogenous silica settling out of the surface water is regenerated in the water column. A further large fraction is returned to the ocean from solution on the sea floor since $1.3\, mg\, SiO_2\, cm^{-2}\, y^{-1}$ is much larger than the amount required to be buried in the sediments in order to maintain a steady state. If $0.3\, mg\, SiO_2\, cm^{-2}\, y^{-1}$, or 10% of the biologically fixed silica, becomes buried (Heath, 1974), then around 35% of the total mass dissolves on the sea floor.

A further demonstration of the efficacy of biological activity in driving the oceanic silicon cycle is provided by an estimate of the residence times of silicon in the surface ocean mixed layer with respect to biological uptake and mixing. The rate constant for biological removal can be derived from (2) as

$$K_b = (C_d/C_s - 1)K_d/f \tag{3}$$

where $f = V_s/V_d$. Taking the surface and deep layer thicknesses given earlier, $f = 0.056$ and $K_b = 0.16\,y^{-1}$. Hence the residence time of silicon in the surface 200 m-thick layer is 6 years with respect to biological removal. Taking the conservation condition for the two water masses, such that:

$$K_s V_s = K_d V_d \tag{4}$$

then the residence time of silicon in the surface layer with respect to removal by mixing is:

$$1/K_s = f/K_d \tag{5}$$

or 60 years. Consequently, the biological mechanism is ten times as effective in removing silicon from the surface layer as is mixing. Heath (1974) has shown, using similar reasoning, that the residence time of silicon with respect to deposition and burial is very much longer than the times derived above and that, consequently, the rate of biological removal will change rapidly in response to changes in the rate of supply or removal due to climatic, tectonic, or oceanographic conditions.

Estimates of the flux of biogenous silica into deep water have recently become available from the use of traps to measure the flux of particulate material to the sea-floor. The most recent measurements give a range of 10–200 $\mu g\,SiO_2\,cm^{-2}\,y^{-1}$ at 5000 m depth from one Pacific and two Atlantic stations (Honjo, 1980). Takehashi and Honjo (1981) give a flux of 42 $\mu g\,cm^{-2}\,y^{-1}$ of *radiolarian* opal of 5288 m depth in the western tropical Atlantic. If we assume that the biological removal of dissolved silicon from the mixed layer in the Atlantic can be estimated by the method of Lerman and Lal (1977), where $C_s = 2$ and $C_d = 50\,\mu mol\,l^{-1}$, it can be shown that these fluxes represent about 20% of the surface biological removal.

The removal of silicon in the oceans on a global scale by the burial of biogenous silica in sediments has been most recently estimated by DeMaster (1981) using radiometric methods to obtain bulk sediment accumulation rates. His figures are summarized in Table 5.2. Approximately 50% of the total removal takes place around Antarctica and fully 70% of the total takes place in the deep sea. The main sites of removal are in high latitudes, with equatorial regions being very minor sinks. Some marginal basins (e.g. the

Gulf of California and the Sea of Okhotsk) do accumulate silica at rates very much higher than those in the deep sea, although, because of their limited sizes, they are not major repositories.

Heath (1974) estimated the total removal of silicon in the ocean and concluded from the available evidence that the inputs are much larger than the known rates of sedimentation. This led him to suggest, following Turekian (1971), that large quantities of silica are accumulating in nearshore environments where its presence is masked by rapidly accumulating terrigenous debris. For example, in the Cascadia Basin off Oregon, opal accumulates at a similar rate (Heath et al., 1976) to that in the Gulf of California, a known sink for silica, although the sediments are not noticeably siliceous. Likewise, opal accumulates at a moderately high rate in the Panama Basin (Moore et al., 1973). DeMaster (1981) obtained a rate of silica accumulation in Long Island Sound of 1.9×10^{10} g y^{-1}. Although this figure is negligible compared with the rates shown in Table 5.2, it serves to illustrate the fact that such nearshore environments are locally important, since this rate is larger than the supply to the Sound by rivers, 5.7×10^9 g SiO_2 y^{-1}. The Sound is a sink for oceanic silicon, as concluded previously by Turekian (1971).

The estuarine removal of silicon given in Table 5.2 is derived from the estimate that roughly 20 % of the riverine flux is extratced during mixing with sea water. The mechanism of removal of silicon during such mixing has been debated extensively, one school arguing that inorganic reactions are involved (Bien et al., 1958; Liss and Spencer, 1970), the other school maintaining that no such removal can be unequivocally detected (Schink, 1967; Fanning and Pilson, 1973; Boyle et al., 1974). On the other hand, Wollast and De Broeu (1971), Milliman and Boyle (1975) and Peterson et al., (1975) have demonstrated that biological removal of silicon is an important process in some estuaries. It may well account for the estimates of silicon removal, from silicon–salinity relationships, claimed from other estuaries. Regardless, however, of the actual mechanisms involved, the estimate of DeMaster (1981) amounts to roughly 20 % of the total removal of silicon in the ocean. The estuarine geochemistry of silicon is discussed in detail in Chapter 3.

5.5   THE SILICON CYCLE IN LAKES

5.5.1   Freshwater Lakes

The concentration of dissolved silicon in fresh waters varies widely, from trace quantities to around 1.7 m mol l^{-1} (Livingston, 1963). It is derived from surficial chemical weathering and from groundwater leaching of rocks. As a result, concentrations are generally higher in ground- and formation-waters (Davis, 1964). The higher concentration of silicon in river waters compared with surface sea water is matched by nitrogen and phosphorus, all three essential plant nutrients in the sea, and in marked contrast to the major ions of sea water.

The dissolved silicon content of fresh waters, at normal ambient values of pH, falls within a fairly narrow range and does not show marked variations with discharge as do other dissolved constituents (Davis, 1964). Edwards and Liss (1973) have suggested that an abiological buffering mechanism exists and that particulate iron and manganese oxyhydroxides may be involved. The dissolved silicon content of soil waters is known to respond rapidly to surface adsorption reactions, most probably involving oxyhydroxides (Beckwith and Reeve, 1964; Miller, 1967). In addition, the cycle of dissolved silicon in the deep layers of some lakes is thought to be affected by seasonal cycles of reductive solution and oxidative precipitation of ferric oxyhydroxides (Mortimer, 1941, 1942; Kato, 1969). Mayer and Gloss (1980) have presented both observational and experimental evidence for this buffering effect, but also suggest that the effect might be caused by equilibration in the soil profile (see Kennedy, 1971).

In spite of these inorganic reactions, it is clear that the dissolved silicon cycle in freshwater lakes is intimately connected with biological activity. Diatoms are often the most important members of the freshwater phytoplankton and in most lakes they are the perenially dominant group (Hutchinson, 1967). The concentration of dissolved silicon varies markedly on a seasonal basis in all temperate-latitude lakes and is a consequence of extraction by planktonic and benthonic diatoms. The long time-series of the observations of Lund (1964) illustrates this cycle well. The vertical distribution of dissolved silicon in lake basins commonly shows surface depletion and deep-water enrichment (Wetzel, 1975), consistent with surface utilization and deep regeneration, as is found in the ocean.

The accumulation of biogenous opal in lake sediments, and the formation of freshwater diatomites, have not received the same attention as this part of the marine cycle. However, lacustrine diatomite deposits are known (Ellenburg and Kuhn, 1967). Tessenow (1966) has shown that the amorphous silica content of the muds of the Grosse Plöner lake in northern Germany is related to water depth, the concentration increasing from less than 1 % by weight to 12 m depth to 15–17 % by weight in samples from 40 m depth. In shallower parts of the lake, the silica decreases with depth in the sediment, whereas in deeper water the content shows a subsurface maximum, reaching a concentration 20 % higher than at the 0–1 cm level. Tessenow is of the opinion that the silicon cycle in the north German lakes is controlled by biological processes and that inorganic reactions are of minor importance.

In some of the North American Great Lakes, estimates of biogenous silica in the sediments have been made by visual diatom cell counts (Duthie and

Sreenivassa, 1971, 1972; Parker and Edgington, 1976); a range from 6.3×10^5 to 3.7×10^7 frustules g^{-1} have been reported for Lake Michigan, with an increase in the direction of deep water. Assuming a silica content of the shells of *Cyclotella* and *Stephanodiscus* as given by Bailey-Watts and Lund (1973) and Bailey-Watts (1976a), this abundance is equivalent to a total biogenous opal content of less than 1 % by weight. The estimate of 3–9 % by weight amorphous silica in Lake Ontario sediments given by Nriagu (1978) is probably an overestimate of biogenous silica because of the method used and the presence of some silica in an amorphous aluminosilicate (see Section 5.6.2). Duthie and Sreenivassa (1971, 1972) find very low concentrations of diatom shells in Lake Ontario sediments.

The concentration of diatom frustules in Lake Michigan sediments decreases exponentially with depth in the sediments by two orders of magnitude (Parker and Edgington, 1976). This is interpreted as a reflection of the solution of the shells within the sediment; shells are severely corroded at depth and pore water dissolved silicon concentrations are significantly higher than in the lake itself (Parker *et al.*, 1977). Estimates of the annual deposition of diatoms in Lake Michigan range from 0.1 to 52×10^4 shells $cm^{-2} y^{-1}$ compared with an annual production of 1.2×10^7 shells $cm^{-2} y^{-1}$. Hence, less than 4 % of the production reaches the underlying sediment (Parker and Edgington, 1976). Conway *et al.* (1977) and Parker *et al.* (1977) further emphasize that the regeneration of silicon from settling diatoms is required to support the annual rate of primary diatom production

TABLE 5.3   Silicon budget for Lake Superior (from Johnson and Eisenreich (1979)).

		10^{11} g SiO$_2$ y^{-1}
1. *Supply*		
River influx		4.2
Atmospheric influx		0.026
Shore-line erosion		0.07
	Total	4.53
2. *Removal*		
River outflow		1.4–1.7
Sedimentation		
Diatoms		0.2–0.3
Adsorption		0.1
Silicate authigenesis		1.9–3.4
	Total	2.2–3.7

in the lake, the annual watershed supply amounting to only 3 % of the annual biological requirement.

Nriagu (1978) has argued that the major fraction of the mass of diatoms produced in Lakes Ontario, Erie, and Superior during the spring bloom do not surive to be buried in the sediment and that most of the silicon is recycled in the lake itself. In the case of Lake Ontario, the amount of silica in the diatom biomass during the spring bloom exceeds the annual supply by a factor of 4.8 so that a large-scale recycling of silicon is required to maintain the production. The magnitude of the flux of dissolved silicon from the sediment pore waters is discussed in Section 5.6.1.

A budget for dissolved silicon in Lake Superior, which illustrates some of these conclusions, is given in Table 5.3.

5.5.2 Saline Lakes

Extensive deposits of chert, consisting of finely crystalline quartz, occur in the Pleistocene deposits of several saline lake beds in East Africa (Hay, 1968; Eugster, 1969). They are intimately associated with two new sodium silicate minerals (Eugster, 1967; Eugster and Jones, 1968) and provide valuable evidence of the silicon cycle in lakes where the compositions of the waters are controlled by processes other than those operating in fresh waters or in the sea. The new silicate phases have also been discovered in lake deposits in Oregon and California (McAtee *et al.*, 1968).

The alkaline brines in the East African lakes are greatly enriched in dissolved silicon (up to 32 mmol l^{-1} in Lake Magadi) because of evaporative concentration and because of polymerization of silicic acid at high pH values. Magadiite ($NaSi_7O_{13}(OH)_3$, $3H_2O$) and kenyaite ($NaSi_{11}O_{20.5}(OH)_4$, $3H_2O$) are precipitated when the lake brines are diluted by freshwater influx (Eugster, 1967, 1969) which lowers the pH and supersaturates the water. Magadiite is evidently precipitated first:

$$7H_4SiO_4 + Na^+ \rightarrow NaSi_7O_{13}(OH)_3 \, 3H_2O + 9H_2O + H^+$$

Gradual leaching of Na from this phase leads to the formation of kenyaite:

$$22NaSi_7O_{13}(OH)_3 \, 3H_2O + 8H^+ \rightarrow 14NaSi_{11}O_{20.5}(OH)_4 \, 3H_2O \\ + 8Na^+ + 33H_2O$$

and this can be carried out in the laboratory.

Chert forms in turn from the silicates by the further removal of Na:

$$NaSi_7O_{13}(OH)_3 \, 3H_2O + H^+ \rightarrow 7SiO_2 + Na^+ + 5H_2O$$

$$NaSi_{11}O_{20.5}(OH)_4 \, 3H_2O + H^+ \rightarrow 11SiO_2 + Na^+ + 5\tfrac{1}{2}H_2O$$

The quartz occurs as thin continuous plates and is finely laminated, as are the precursor minerals. Oxygen isotope evidence (O'Neil and Hay, 1973) indicates that the cherts formed in contact with lake water of widely varying salinity and sodium and hydrogen ion activities.

The discovery of extensive beds of chert in a lacustrine environment suggests that inorganic precipitation of silica can be important in some environments where organic activity does not regulate the concentration of dissolved silicon in the water. This happens in highly saline environments and these in turn are restricted to lakes. Eugster (1969) and Eugster and Chou (1973) have suggested that many Pre-Cambrian bedded cherts were also formed in extensive alkaline lakes by the mechanism described here.

5.6   DIAGENESIS OF SILICA

5.6.1   Solution of Silica

From the previous discussions concerning the budget of silicon in the ocean and in freshwater lakes, it is clear that the solution of biogenous silica both in the water column and in sediments is the principal mechanism for recycling dissolved silicon through these systems. Within marine and lacustrine sediments, silica is evidently a labile component and leads to the extensive series of alterations which ultimately produce chert.

The high concentration of dissolved silicon in the pore waters of marine and lacustrine sediments is a direct reflection of this recycling. The available data (Siever et al., 1965; Tessenow, 1966; Fanning and Schink, 1969; Bischoff and Ku, 1970, 1971; Fanning and Pilson, 1971; Bischoff and Sayles, 1972; Heath and Dymond, 1973; Hurd, 1973; Sholkovitz, 1973; Weiler, 1973; Nriagu, 1978; Johnson and Eisenreich, 1979) show that dissolved silicon increases rapidly in concentration in the uppermost layers of the sediment and reaches a constant value some tens of centimetres in the sediment (Fig. 5.4).

The increased concentrations of dissolved silicon in the pore waters are brought about by solution of planktonic shells and most of this occurs close to the sediment surface. The extent of such solution is determined by the longer residence time of settled particles at the surface compared with the transit time through the water column, and possibly also by benthic feeding activities (Tessenow, 1966). Schink et al. (1974) and Johnson and Eisenreich (1979) have shown that dissolved silicon concentrations are highest where the sediments contain more biogenous silica and vice versa. The concentration levels reach $500 \, \mu mol \, l^{-1}$ which are less than equilibrium solubility levels with respect to amorphous silica, about $1100 \, \mu mol \, l^{-1}$ at $2°C$ (Wollast, 1974), and five to seven times higher than the solubility level with respect to quartz (Siever, 1957).

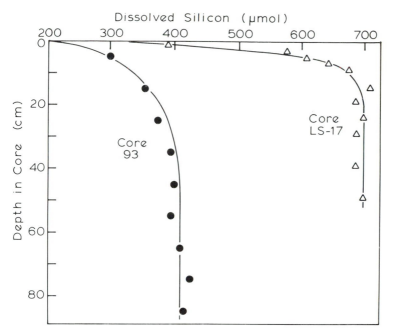

FIG. 5.4   Examples of concentration depth profiles of dissolved silicon in the pore waters of siliceous clay (Core 93; Hurd, 1972) and Lake Superior mud (Core LS-17; Nriagu, 1978).

Schink *et al.* (1974) mapped the variation in the concentration of dissolved silicon in the pore waters of Atlantic Ocean sediments (Fig. 5.5) and showed that the concentrations are highest around the margins of the North Atlantic gyre. Such a variation is consistent with the distribution of biogenous silica in the sediments, expressed both as total amorphous silica (Yemel' Yanov, 1975) and as numbers of radiolarian shells (Goll and Bjørkland, 1971, 1974). This distribution in turn is brought about by high surface production in marginal areas and a negligible flux of biogenous debris to the bottom in the central gyres.

The shape of the concentration depth profile of a dissolved constituent in the pore waters of a sediment is the result of a balance between influx, dissolution, diffusion and advection. Although the movement of water as a result of compaction is unimportant where the sediment accumulation rate is low, physical disturbance of the near-surface sediments by burrowing organisms is nevertheless important (Schink *et al.*, 1975). Mathematical models of the concentration-depth profiles of dissolved silicon have been developed by Anikouchine (1967), Hurd (1973), Schink *et al.*, (1975), Schink and Guinasso (1977), and Wong and Grosch (1978) in order to attempt to separate the effects of physical, chemical, and biological processes in producing the observed distributions. Schink *et al.* (1975) showed that the prifile is most sensitive to biological mixing rates in the surface sediments, input of particulate biogenous silica and the solution rate of the silica.

A scale length for the distribution of dissolved silicon in pore waters is given by the ratio of the biological mixing rate to the solution rate constant (D_B/K_B). In order to fit observed profiles with their model, Schink *et al.* (1975) assumed high values of D_B relative to K_B which implied surprisingly high biological mixing rates in pelagic sediments and that the solution rate of the particles was lower than expected. This latter conclusion may be supported by diagenetic processes which alter the physical properties of the opal which would retard solution (see Section 5.6.2.). Guinasso and Schink (1975) found that values of the biological mixing rate varied widely in pelagic sediments, from 1 to 1000 cm^2 K y^{-1}.

Wong and Grosch (1978), in a re-evaluation of the mathematical aspects of the model of Schink *et al.* (1975), reaffirmed the importance of the ratio D_B/K_B in determining the shape of the concentration/depth profile of

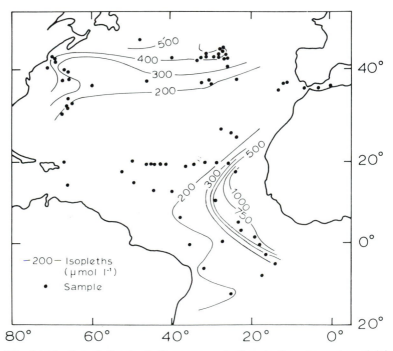

FIG. 5.5 Distribution of dissolved silicon concentrations in the pore waters of Atlantic Ocean sediments. Values are averages of the more or less constant values attained in the upper 100–200 cm of the sediment cores. Modified from Schink *et al.* (1974).

dissolved silicon in the near-surface zone of pelagic sediments and concluded that the steady-state concentration at depth (see Fig. 5.4) is controlled primarily by the flux of silica particles to the sediment. Moreover, where all dissolvable particles dissolve at depth, the flux of silicon across the sediment surface is governed by the sediment porosity and the concentrations of silicon in the bottom water and the deep pore waters.

The flux of dissolved silicon across the sediment/water interface has been estimated by applying the relationship:

$$F = \phi D \frac{dc}{dZ_{z=0}} \tag{5}$$

where

F = flux (mol cm^{-2} s^{-1})
ϕ = porosity
D = diffusion coefficient of dissolved silicon ($\sim 4 \times 10^{-6}$ cm^2 s^{-1})
C = concentration of dissolved silicon (mol cm^{-3})
Z = depth (cm)

to many concentration/depth profiles.

The derived fluxes, averaged over the entire ocean floor, range between 0.6 and 7.9×10^{14} g SiO$_2$ y^{-1}. The amount of recycled dissolved silicon from the pore waters of marine sediments is of the same magnitude as that supplied by run-off.

5.6.2 Early Diagenetic Reactions

As noted in Section 5.6.1, the concentrations of dissolved silicon in the pore waters of sediments do not reach equilibrium solubility values with respect to amorphous silica, even in highly siliceous sediments. This observation has led to speculations about the mechanisms which control the concentration levels.

The ageing or incipient ordering of amorphous silica during burial will influence the solution rate. Such effects have been reported for diatoms (Rogall, 1939) and it is known that the ease of conversion of opal to cristobalite by heating decreases with age (Heath, 1968). In addition, Kamatani (1974) has shown that diatom opal dissolves more slowly when it has been heated (from 100–900°C), the solution behaviour being different for different species. Although no evidence for a change in the crystallinity of the heated opal can be detected by conventional X-ray diffraction methods, the adsorbed water is permanently lost from the samples heated above 100°C.

Hurd and Theyer (1975, 1977) have examined the changes in the properties of radiolaria and sponge spicules with age in Pacific sediments. They found that the solubility and the water content decrease and the density increases in older samples. Concomittantly, the specific surface area of the opal has decreased by two orders of magnitude over the past 40 million years. No significant change in the sizes of the fine particles making up the opal could be seen, consistent with earlier observations of Heath and Moberley (1971) and von Rad and Rösch (1974).

The reaction of dissolved silicon released from biogenous opal with other components of the sediment is an additional mechanism which could control the concentration in the pore solution. The evidence for such reactions is contradictory. Bischoff and Sayles (1972), Hurd (1973), Heath and Dymond (1973), and Nriagu (1978) have pointed out that the concentrations observed in many sediments are close to the equilibrium levels with respect to several ideal silicate mineral phases. Nriagu (1978) suggests that it is more likely that the silicon concentrations in the sediments of Lakes Ontario, Erie, and Superior are controlled by the precipitation of amorphous ferroaluminium silicates from dissolving silica and ferric and aluminium hydroxides, the latter constituents presumably being derived from the breakdown of more crystalline materials.

In opposition to this conclusion, Johnson and Eisenreich (1979) have argued that a smectite phase is forming in the sediments of Lake Superior and that this reaction is required to control the observed level of dissolved silicon in the pore waters. They show an increasing smectite content with depth in some sediment cores from the lake, but this is evidently not shown by all cores and it is not clear if it also reflects to some degree a change in the sedimentary regime of the lake basin over the recent past. Fanning and Schink (1969) had earlier concluded that clay mineral reactions were not capable of explaining the observed distributions of dissolved silicon in the pore waters and the observations of Schink et al. (1974) are consistent with this. A resolution of this problem must await the development of new techniques and approaches for the study of sediment–water interactions.

5.6.3 Recrystallization of Silica

The pathways and the products of the recrystallization of biogenous opal have become reasonably well known from the results of mineralogical and petrographic studies of rocks and sediments recovered by the Deep Sea Drilling Project. In addition, new studies of the Miocene Monterey Formation of California have provided confirmatory evidence for the mode of formation of chert from biogenous opal (see Section 5.6.5). These processes were previously poorly understood from work on ancient chert sequences.

Early discoveries of partially recrystallized silica in marine sediments have been summarized by Calvert (1974). Poorly ordered cristobalite was discovered in claystones from deep-sea drill cores (Calvert, 1971a,b; Heath and Moberly, 1971) shortly after this phase was identified in Monterey Formation porcelanites (Ernst and Calvert, 1969). The cristobalite, or opal-CT in the terminology of Jones and Segnit (1971), was considered to be an intermediate phase in the conversion of amorphous silica to quartz, and subsequent work has largely confirmed this conjecture.

The opal-CT has a characteristic X-ray diffraction pattern (Fig. 5.6) showing two broad reflections at 405–410 and 250 pm and a subsidiary reflection at 425–435 pm. The reflections in the 405–435 pm region in some opals have previously been assigned to β-cristobalite (Greigg, 1932; Levin and Ott, 1933; Dwyer and Mellor, 1933; Buerger, 1951; Omure *et al.*, 1956; Buerger and Shoemaker, 1972), although it is now known, thanks to the extensive work of Flörke (1955a,b), that the patterns are produced by undimensionally disordered, low-temperature cristobalite, where random stacking of the silica tetrahedra produces reflections of tridymite. Such material, named *lussatite* by Mallard (1890), occurs in a wide variety of silica deposits formed at low temperature, such as opals, glasses, and siliceous concretions (Flörke, 1955a; Swineford and Franks, 1959; Gade *et al.*, 1963; Garavelli, 1964; Jones *et al.*, 1964; Mizutani, 1966) and in bentonites and siliceous claystones (Gruner, 1940; Heron *et al.*, 1965).

Further evidence for the stacking disorder in opal-CT is provided by the (101) spacing of 411 pm compared with a value of 404 pm in well-crystallized cristobalite. Heating opal at 1050–1100°C results in a decrease in the spacing from 411–404 pm, the product being structurally identical to α-cristobalite.

Low-temperature hydrous silicas have a wide range of structures, which have been examined by Jones *et al.* (1964). They range from more or less amorphous varieties to well-crystallized cristobalite. The classification scheme of Jones and Segnit (1971) arising from this work is shown in Table 5.4. Examination of the X-ray data of deep-sea porcelanites shows that they all fall into the opal-CT class.

An alternative interpretation of the structure of opal-CT has been offered by Wilson *et al.* (1974) based on electron diffraction and infra-red absorption data. They consider the phase to be tridymite, since it has a hexagonal morphology and diffraction pattern, and suggest that the distinctive X-ray diffraction pattern is caused by random transverse displacements of silica tetrahedra normal to the C-axis rather than irregular stacking of cristobalitic and tridymitic sheets as proposed by Flörke (1955a). This explanation has been rejected by Jones and Segnit (1975) who have argued that the hexagonal morphology and electron diffraction pattern are equally well produced by disordered cristobalite.

Individual crystals of silica lining a radiolarian test in an Eocene claystone

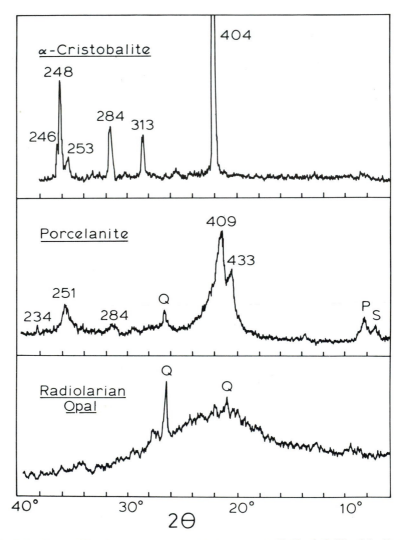

FIG. 5.6 X-ray diffraction patterns of radiolarian opal (Gulf of California), Eocene porcelanite (opal-CT) from DSDP site 12 in the North Atlantic and well-crystallized α-cristobalite (made by heating silica gel at 1350°C for 4 hours). P, polygorskite; S, sepiolite; Q, quartz. Reflections with spacings (pm) indicated are due to typical α-cristobalite and opal-CT.

FIG. 5.7 Scanning electron photomicrographs of opal-CT. (A) Cluster of lepispheres in Campanian-Maestrichtian silicified foraminiferal marlstone, DSDP site 144, Core 3, Section 2, 103–104 cm, north-west Atlantic Ocean. (B) Opal-CT lepispheres from Middle Eocene Zeolitic clay and porcelanite, DSDP Site 367. Core 10, Section 1, 126–128 cm, north-east Atlantic Ocean. Note bladed crystals interpenetrating at around 70° (photomicrographs courtesy of U. Von Rad and H. Rösch).

from the Atlantic have been identified as high temperature forms of cristobalite and tridymite (Klasik, 1975) based on morphological data. Such crystals may in fact be identical to opal-CT in view of the previous discussion about the morphology of opal-CT and the detailed analysis of the habit of opal-CT crystals by Flörke *et al.* (1975, 1976), together with the fact that all well-crystallized tridymite crystals found in nature are inversion pseudo-morphs after high tridymite (Calvert, 1974).

TABLE 5.4   Classification of natural hydrous silicas (from Jones and Segnit (1971)).

Variety	X-ray diffraction data	Interpretation	Occurrence
Opal-A	Broad diffuse band at 410 pm; low-intensity bands at 200, 150, and 120 pm	Highly disordered; nearly amorphous	Precious opals; potch opals; diatom and radiolarian shells; hyalite; geyserite
Opal-CT	Broad reflections at 410 and 250 pm; subsidiary reflection at 425–435 pm	Unidimensionally dis-ordered cristobalite-tridymite	Common opals; tripoli; opoka; porcelanites
Opal-C	Sharp reflections corre-sponding to well-crystallized α-cristobalite	Well-ordered α-cristo-balite	Opals associated with lava flows

Opal-CT which grows in cavities in deep-sea and terrestrial porcelanites occurs as spheres, of the order 10–15 μm in diameter, composed of finely bladed crystals 30–50 nm thick (Fig. 5.7). The spherical aggregates were termed *lepispheres* by Wise and Kelts (1972) and have since been described from many primary and replacement silicas. Opal-CT lepispheres have also been synthesized in the laboratory (Oehler, 1973; Flörke *et al.*, 1975; Kastner *et al.*, 1977). The bladed habit is evidently responsible for the lack of certain cristobalite reflections in X-ray diffraction patterns of the phase, a feature noted in earlier discussions of its structure (Flörke, 1955a,b). The crystals, which interpenetrate at 70–71° (see Fig. 5.7), are crudely hexagonal and represent hybrid crystals of cristobalite (111) and tridymite (0001), together with cristobalite twins (Flörke *et al.*, 1976).

5.6.4   Formation of Chert

The evidence gathered from studies of siliceous sediments and rocks from the deep sea and of siliceous formations in land sequences is consistent with a two-stage formation of chert. Biogenous silica dissolves in sediments and crystallizes as opal-CT before it, in turn, dissolves and is reprecipitated as fine-grained quartz. An intermediate cristobalite phase has been frequently observed in experiments on the recrystallization of amorphous silica (Carr and Fyfe, 1958; Campbell and Fyfe, 1960; Heydemann, 1964; Mizutani, 1966; Sieffert and Wey, 1967). Bettermann and Liebau (1975) have shown that a range of intermediates (SiO_2–X, keatite and cristobalite) can also be formed in a well-defined sequence depending on the pressure and hydroxyl ion activity of the solution. At high pressure (300 MPa), or at high OH^- activities, such an intermediate phase does not form and quartz precipitates directly.

The evidence for the two-stage formation of chert comes largely from petrographic and mineralogical observations. In deep-sea sediments and rocks, porcelanites, consisting of opal-CT and ranging in age from Pliocene to Late Jurassic occur in a variety of lithologies, including calcareous oozes and chalks, siliceous oozes, and terrigenous and zeolitic claystones (Calvert, 1977). In general, they are younger than sequences of chert at any one drill site, although there is no clear age relationship between the two rock types when different sites are compared. This is explained by the difference in *in situ* temperatures at the various sites (Heath and Moberly, 1971).

In the northern Pacific belt of diatomaceous ooze, the conversion of biogenous silica to opal-CT occurs at around 600 m below the sea-floor (Hein *et al.*, 1978). The conversion of unconsolidated sediment to silica-cemented porcelanites produces a regional sub-sea floor acoustic reflector which "mirrors" the sea-floor topography. Using the temperature gradients measured in the deep-sea drill holes, it can be shown that opal-CT begins to form at around 35°C, consistent with the temperature of transformation in the Monterey Formation (see Section 5.6.5). The conversion horizon, the acoustic reflector, is evidently an isothermal surface, its topography depending on the depth of burial and the local geothermal gradient (Scholl and Creager, 1973). This "diagenetic front" moves upwards with continued sedimentation.

The depth and age distribution of porcelanites and cherts in the Atlantic Ocean have been reviewed by von Rad and Rösch (1974) and Riech and von Rad (1979). Figure 5.8 shows that although there is a considerable degree of overlap, quartz is abundant in older and more deeply-buried rocks, unaltered biogenous opal occurs in the shallowest and youngest sediments, while opal-CT occupies an intermediate position. Riech and von Rad (1979) maintain that the evidence indicates that quartzitic cherts always form from an opal-CT precursor.

A relationship between the mineralogy of deep-sea siliceous rocks and the associated lithology has been described by Lancelot (1973) and Greenwood (1973). In sequences of equivalent age, quartzitic cherts occur predominantly in calcareous sediments while porcelanites (containing opal-CT) occur in non-calcareous sequences. Lancelot (1973) thought that this would be consistent with the requirement of "foreign cations" for the formation of disordered cristobalite-tridymite (Flörke, 1955a). Quartz would form preferentially where such impurities were absent or in low concentration.

Kastner *et al.* (1977) have shown experimentally that opal-CT is formed from biogenous opal by solution and reprecipitation. The rate of precipitation depends on temperature and time, and also on the composition of the solution and the mineralogy of the host sediments. Evidently, the presence of magnesium ions and a source of hydroxyl ions are necessary for the conversion. Under hydrothermal conditions, both magnesium ions and alkalinity are consumed during the formation of opal-CT and the presence of magnesium hydroxide enhances the formation of silica lepispheres. The reduction in the alkalinity of the solution phase promotes the dissolution of calcite (from biogenous tests) thereby providing a continuous source of alkalinity for the reaction The role of magnesium in the precipitation of silica has previously been identified by Iler (1955). In experiments where calcite is replaced by smectite, Kastner *et al.* (1977) found that opal-CT does not form and explained this by the inadequate supply of alkalinity and the adsorption of magnesium by the clay. In addition, Kastner and Keene (1975) have

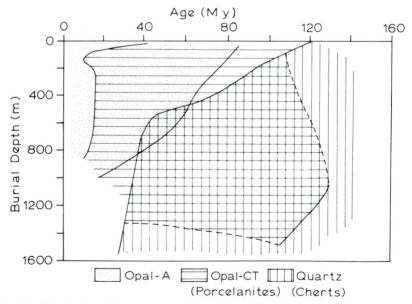

FIG. 5.8   Distribution of cherts and porcelanites in the Atlantic Ocean with respect to age and depth of burial. Modified from Riech and Von Rad (1979).

suggested that direct precipitation of quartz also occurs, as indicated by the experiments of MacKenzie and Gees (1971), when the concentration of silica in the solution is less than the equilibrium solubility values for amorphous silica. Both of these mechanisms would explain the observations of Lancelot (1973) and Greenwood (1973).

The formation of quartz from opal-CT has been studied experimentally by Ernst and Calvert (1969) using Monterey Formation porcelanite. From the results of hydrothermal experiments at 300°, 400°, and 500°C at 200 MPa pressure, they concluded that the mechanism was a solid-state inversion of opal-CT to quartz, the rate depending entirely on the temperature. A representative set of results is shown in Fig. 5.9. Although petrographic evidence cited by Heath and Moberly (1971) and Heath (1973) is consistent with this hypothesis, a reinterpretation of the experimental results by Stein and Kirkpatrick (1976) suggests a different mechanism. The experimental run products were shown to contain euhedral quartz crystals that could only

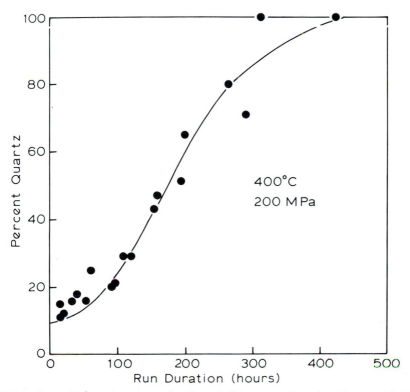

FIG 5.9   Rate of conversion of porcelanitic opal-CT to quartz. Data from Ernst and Calvert (1969); curve fitted as recommended by Stein and Kirkpatrick (1976).

have been formed by precipitation from solution. This is also the mechanism suggested by Mizutani (1966) from experiments on the crystallization of silicic acid at high pH. Note that Riech and von Rad (1979) show some evidence for the solid state inversion of opal-CT to quartz in cherts and porcelanites from the Atlantic Ocean.

5.6.5   The Monterey Formation

The Miocene Monterey Formation of California is a distinctive, widely distributed formation consisting of a number of different siliceous rock types. The classic account of the rocks by Bramlette (1946) contains full descriptions of these lithologies, their stratigraphic relationships, and their modes of formation. An up-to-date summary of modern work on the formation is given in Garrison and Douglas (1981). The siliceous rocks reach thicknesses of more than 2 km in central California. In general, Bramlette established that the lower part of the formation consists of quartzitic chert and porcelanite, whereas unaltered diatomite increases in volume in the upper part of the formation. This distribution, plus the wealth of petrographic detail assembled by Bramlette, suggested to him that the cherty rocks formed by the alteration of diatomite through burial metamorphism. Ernst and Calvert (1969) re-iterated this conclusion and further suggested that opal-CT (cristobalite) is formed as an intermediate phase in the conversion and showed that the so-called porcelanites in the middle part of the formation characteristically consist of this phase. The conversion is by way of solution and reprecipitation, and this is confirmed by the discovery of opal-CT lepispheres in cavities in some porcelanites (Oehler, 1975).

Further examination of the mineralogy of the siliceous rocks by Murata and Nagata (1974) and Murata and Larsen (1975) has added considerable detail to this scheme of alteration. Figure 5.10 shows the distribution of siliceous rock types and their mineralogies in a 2 km-thick, continuously exposed section in central California. The upper part of the formation contains abundant unaltered diatoms, the intermediate beds, 1200 m thick, contain abundant opal-CT (Murata and co-workers refer to the silica phase as cristobalite, although their X-ray diffraction patterns are identical to those of opal-CT), and the lower part of the formation, 1000 m thick, consists of quartzitic cherts and porcelanites.

Murata and Nagata (1974) further established that the structure of the opal-CT, as determined by X-ray diffraction, also changes systematically through the middle part of the formation (Fig. 5.10). In the younger beds, the d(101) spacing of the opal-CT lies around 411 pm and decreases to around 404 pm (a value typical of α-cristobalite) at the base of the opal-CT-bearing section. This change is interpreted as a progressive ordering of the opal-CT brought about by increasing pressure and temperature. This is accompanied by a progressive sharpening of the principal d(101) reflection, a change also noted by Oehler (1973). Mitsui and Taguchi (1977) have described similar changes in the ordering of opal-CT in the Miocene Wakkanai Formation in Japan, von Rad *et al.* (1977) have found progressive ordering of opal-CT in some North Atlantic porcelanites and Mizutani (1977) has prepared opal-CT hydrothermally, having d(101) spacings ranging from 405–410 pm depending on the reaction temperature.

Murata and Larson (1975) estimated the burial temperatures of the Monterey section in central California and concluded that the opal–opal-CT and opal-CT–quartz transformations took place at around 50°C and 110°C, respectively. Confirmatory evidence for these temperature ranges is provided by oxygen isotope ratios in the various silica minerals which yield temperatures of 48 ± 8 and 79 ± 2°C for the two transformations (Murata *et al.*, 1977). The changes in the isotope ratios occur abruptly at the transitions from opal to opal-CT and from opal-CT to quartz, implying that re-equilibration of the oxygen isotopes took place during the solution-

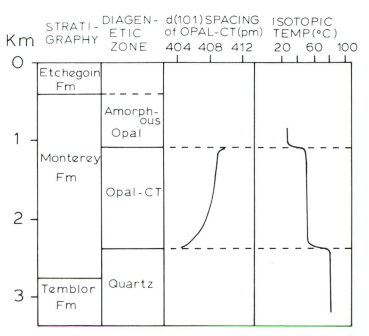

FIG. 5.10. Stratigraphic distribution of diagenetic silica phases in the Monterey Formation at Chico Martinez Creek, California. Modified from Murata and Larsen (1975) and Murata *et al.* (1977).

reprecipitation reactions and solid state ordering of the opal-CT was paramount in the central part of the formation.

Murata and Randall (1975) and Murata *et al.* (1979) have extended the information on the progressive change in the mineralogy of the silica phases in the Monterey Formation to the solution of some structural problems in the formation, in particular the delineation of a folded structure in the Monterey Shale and the location of two possible buried serpentine bodies which have warmed the overlying shale and accelerated the conversion of opal-CT to quartz.

The diagenetic sequence of mineral transformations discovered in the Monterey Formation clearly shows the rather complex way in which quartzitic chert forms from biogenous opal and corroborates the suggested sequence of alterations from scattered drill cores in late Mesozoic and Cenozoic rocks in the deep sea (see summaries in Keene, 1976 and Riech and von Rad, 1979). The same sequence has been identified in other bedded and nodular cherts (Scheere and Laurent, 1970; Laurent and Scheere, 1971; Inoue, 1973; Robertson, 1977). The mechanism has been revealed because these rock sequences are youthful, and because there has not been sufficient time and burial (hence higher temperature) to change the entire mass of silica into chert. Older cherts have lost all traces of any precursor minerals and their origin has remained obscure (Ernst and Calvert, 1969).

5.6.6   Alternative Sources of Silica

In addition to the clear evidence that biological agencies have been responsible for concentrating large amounts of silica in sedimentary formations, volcanic sources have also been invoked to explain such deposits. This explanation requires a supply of dissolved silicon from volcanic activity to the point where inorganic precipitation takes place. Davis (1918) and Taliaferro (1933) accepted this mechanism for the Jurassic Franciscan Formation of California where volcanic rocks do occur in intimate association with cherts. Moreover, the common world-wide association of radiolarian cherts and ophiolites is often considered to be a genetic one, where silica is derived from basic magmatic activity in rather deep basins of deposition (Grunau, 1965). The alteration of volcanic ash has also been considered a source of silica for the silicification of wood (Murata, 1940), this association being suggested as an indicator for the presence of volcanic material in sedimentary deposits. As Pettijohn (1957) has pointed out, however, many thickly-bedded cherts have no known association with volcanic rocks and consequently volcanism cannot be accepted as a general mechanism for siliceous sediment accumulation.

Many altered pyroclastic rocks, including bentonites (Hewett, 1917), contain opal-CT, montmorillonite, and one or more zeolites (Bramlette and Posnjak, 1933; Deffeyes, 1959; Hay, 1966; Iijima and Utada, 1966; Reynolds and Anderson, 1967; Reynolds, 1970). Gruner (1940) considered the origin of cristobalite (opal-CT) in such rocks, discounted a high temperature formation, but favoured its formation from the alteration of volcanish ash. He further noted that the opal-CT represented an intermediate stage in the alteration, the silica eventually altering to quartz.

On the basis of the mineralogical associations of opal-CT in ancient bentonites and claystones which were considered to represent altered volcanic rocks (Reynolds, 1970), Calvert (1971b) suggested that some of the cherts and porcelanites in the North Atlantic, also occurring in montmorillonitic and zeolitic clays containing abundant sepiolite and palygorskite, were also formed from silica derived from hydrothermal sources or from altered volcanic debris. Gibson and Towe (1971) and Mattson and Passagno (1971) went further and contended that the widespread cherts in the North Atlantic, identified with a prominent sub-bottom acoustic reflector, recorded a very widespread volcanic event in Eocene time, the source of the debris being identified as the Caribbean. Many of the Eocene claystones in the southeastern United States were also considered to be altered volcanic ashes derived from this activity (Gibson and Towe, 1971).

Heath (1973) has presented a series of arguments for the derivation of the silica in deep-sea porcelanites and cherts predominantly from biological debris, in spite of similarities between some porcelanites and bentonites. In a re-examination of the composition of many of the Tertiary opaline claystones of the United States, Wise and Weaver (1973; 1974) and Weaver and Wise (1974) have emphatically rejected a volcanic origin, finding abundant remains of microscopic siliceous organisms in many of the rocks. They consider the claystones to be altered diatomites.

5.7 THE GEOCHEMICAL BALANCE OVER GEOLOGICAL TIME

The identification of the mechanisms which have controlled the chemical composition of sea water over geological time involves questions concerned with the buffering mechanism, the constancy of the composition of sea water, and the nature of the sedimentary products. Simply stated, the problem is that, assuming a steady state, the oceans receive considerably more Na^+, K^+, Mg^{2+}, HCO_3^-, and $Si(OH)_4$ from river run-off that can be accounted for by known removal mechanisms (Mackenzie and Garrels, 1966a). Garrels and Mackenzie (1967) showed that simple isothermal evaporation of spring and lake waters isolated from solid weathering products produces water of the composition of a soda lake high in alkali cations, low in alkaline-earth cations and with a pH of about 10. In order to account for the difference between such waters and sea water, Mackenzie and Garrels (1966a,b), following Sillén (1961), postulated that silicate mineral equilibria in the oceans represented the buffering mechanism and were responsible for the maintenance of the unique composition of sea water. Note the alternative view of this problem of Pytkowicz (1967; 1975).

The basic reaction involved in this equilibration, as proposed by Mackenzie and Garrels, is the reconstitution of silicate structures, which have been degraded in the weathering profile and stripped of cations and silica, by the reaction of so-called amorphous aluminosilicates with bicarbonate, silica, and cations in sea water to form more crystalline aluminosilicates and with the release of CO_2. The reaction can be written:

$$\text{amorphous-Al-silicate} + HCO_3^- + SiO_2 + \text{cations}$$
$$= \text{cation} - Al - \text{silicate} + CO_2 + H_2O$$

In these reactions, silica is clearly a key reactant and the implication is that this equilibration also controls the dissolved silicon concentration in the ocean. A series of experiments on the release and uptake of silicon from and by silicate minerals (Garrels, 1965; Mackenzie and Garrels, 1965; Mackenzie et al., 1967; Siever, 1968b; Siever and Woodford, 1973; Lerman and Mackenzie, 1975) appears to show that re-equilibration of dissolved silica and clay mineral particles does take place in the laboratory since dissolved silicon is released from the clays when suspended in silicon-deficient sea water and is removed from solution when the clays are suspended in silicon-enriched sea water. This evidence has been taken to indicate the efficacy of wide-spread silicate reactions in the modern ocean which bring about profound changes in the compositions of both solid mineral phases and the water in contact with such phases. Sophisticated equilibrium models of silicate sea water have been developed (Helgeson and Mackenzie, 1970) and roundly criticized (Perry, 1971).

Recent experiments (Wollast and De Broeu, 1971; Siever and Woodford, 1973) have established that the "equilibrium" dissolved silicon values attained are always much higher than the concentration of dissolved silicon in most sea waters, except perhaps in the deep North Pacific. Moreover, the reactions producing the solution/precipitation effect are probably surface reactions, these varying substantially depending on the initial treatment of the solid products used in the experiments. Equilibrium in a restricted chemical sense is in all probability not attained, and the conventional thermodynamic approaches are inadequate to describe the complex mineral mixtures, including metastable phases, of natural sediments. Kinetic considerations are probably required (Wollast, 1974). The few attempts to detect silicate reconstitution in modern marine sediments (for example

Russell, 1970), have failed, simple cation exchange being the only changes observed.

The discussion in Section 5.3 has shown that biological processes control the cycling of silicon in the modern ocean, and as long as silica-secreting organisms have been present in marine environments (Harper and Knoll, 1975), it seems reasonable to assume that such a proven mechanism has operated in contemporary oceans of the past. Nevertheless, the dilemma addressed by Mackenzie and Garrels (1966a,b) remains because the renewal times of many of the major constituents of sea water are very short. The contemporary view of the dilemma is that silicate reactions take place in buried sediments (Calvert, 1968; Siever, 1968a; Siever and Woodford, 1973) as is illustrated by the burial diagenesis of clay minerals in thick sedimentary sections (Perry and Hower, 1970). Whether the products of such reactions leak into the ocean at a low level and thereby influence its composition is an important question requiring much further study. Furthermore, hydrothermal reaction in oceanic crustal rocks, the scale of which is illustrated by the observations of Edmond et al. (1979), is also potentially capable of exerting an important control on the composition of sea water, although silica in particular appears to require special consideration (Maynard, 1976).

REFERENCES

Aleksina, I. A. (1960). *Dokl. Akad. Nauk. S.S.S.R.*, **139**, 624.

Anikouchine, W. A. (1967). *J. Geophys. Res.*, **72**, 505.

Bailey-Watts, A. E. (1976). *Freshwater Biology*, **6**, 69.

Bailey-Watts, A. E. and Lund, J. W. G. (1973). *Biol. J. Linn. Soc.* **5**, 235.

Bardossy, G., Konda, I., Rapp-Shik, S., and Tolnai, V. (1965). *In* "Problems of Geochemistry" (Khitarov, N. I., ed.), Izdat. "Nauka", Moscow. Israel Prog. Scientific Translations (1969).

Beckwith, R. S. and Reeve, R. (1926), *Aust. J. Soil Res.*, **1**, 157.

Bell, R. J. and Dean, P. (1972) *Phil. Mag.*, **25**, 1381.

Berger, W. H. (1970). *Bull. Geol. Soc. Amer.*, **81**, 1385.

Betterman, P. and Liebau, F. (1975). *Contrib. Mineral Petrol.*, **53**, 25.

Bezrukov, P. L. (1955). *Dokl. Akad. Nauk S.S.S.R.*, **103**, 473.

Bezrukov, P. L. (1960). *Rept. Int. Geol. Cong., Copenhagen.*, Part X, 39.

Bien, G. S., Contois, D. E., and Thomas, W. H. (1958). *Geochim. Cosmochim. Acta.*, **14**, 35.

Bischoff, J. L. and Ku, T. L. (1970). *J. Sedim. Petrol.*, **40**, 960.

Bischoff, J. L. and Ku, T. L. (1971). *J. Sedim. Petrol.*, **41**, 1008.

Bischoff, J. L. and Sayles, F. L. (1973). *J. Sedim. Petrol.*, **42**, 711.

Bogoyavlenskiy, A. N. (1967). *Intern. Geol. Rev.*, **9**, 133.

Booth, B. C. and Harrison, P. J. (1979). *J. Phycol.*, **15**, 326.

Boström, K., Kraemer, T., and Gartner, S. (1973). *Chemical Geol.*, **11**, 123.

Boyle, E., Collier, R., and Dengler, A. (1974). *Geochim. Cosmochim. Acta.* **38**, 1719.

Bramlette, M. N. and Posnjak, E. (1933). *Amer. Mineral.*, **18**, 157–171.
Bramlette, M. N. (1946). *U.S. Geol. Survey. Prof. Paper.*, 212, 57 pp.
Brandenberger, E. and Frey-Wyssling, A. (1947). *Experimentia*, **3**, 492.
Brewster, N. A. (1980). *Bull. Geol. Soc. Amer.*, **91**, 337–347.
Bruevich, S. V. (1953). *Izv. Akad. Nauk. S.S.S.R. Ser. Geol.*, **4**, 67.
Buerger, M. J. (1951). *In* "Phase Transformation in Solids" (Smoluchowski, R., Mayer, J. E., and Weyl, W. A., eds.). J. Wiley, New York.
Buerger, M. J. and Shoemaker, G. L. (1972). *Proc. Nat. Acad. Sci. USA*, **69**, 3225.
Burton, J. D. and Liss, P. S. (1973). *Geochim. Cosmochim. Acta.*, **37**, 1761.
Calvert, S. E. (1966). *Bull. Geol. Soc. Amer.*, **77**, 569.
Calvert, S. E. (1968). *Nature*, **219**, 919.
Calvert, S. E. (1971a). *Nature.*, **234**, 133.
Calvert, S. E. (1971b). *Contrib. Mineral. Petrol.*, **33**, 273.
Calvert, S. E. (1974). *In* "Pelagic Sediments: On Land and Under the Sea." (Hsü, K. J. and Jenkyns, H. C., eds.). *Spec. Publ. Assoc. Sediment.*, **1**, 273.
Calvert, S. E. (1977). *Phil. Trans. R. Soc. Lond.*, *A.* **286**, 239.
Calvert, S. E. and Price, N. B. (1971). *In* "The Geology of the East Atlantic Continental Margin". (Delany, F. M., ed.). Inst. Geol. Sci. Rept. **70/16**, London.
Campbell, A. S. and Fyfe, W. S. (1960). *Amer. Mineral.*, **45**, 464.
Carr, R. M. and Fyfe, W. S. (1958). *Amer. Mineral.*, **43**, 908.
Cartz, L. (1964). *Z. Kristall.*, **120**, 241.
Casey, R. E. (1971). *In* "The Micropalaeonotology of Oceans" (Funnell, B. M. and Riedel, W. R., eds.). Cambridge Univ. Press, Cambridge.
Chester, R. and Elderfield, H. (1968). *Geochim. Cosmochim. Acta.*, **32**, 1128.
Coes, L. (1953). *Science.*, **118**, 131.
Conway, H. L., Parker, J. I., Yaguchi, E. M., and Mellinger, D. L. (1977). *J. Fish. Res. Bd. Canada*, **34**, 537.
Coombs, J., Spanis, C., and Volcani, B. E. (1967a). *Plant Physiol.*, **42**, 1607.
Coombs, J., Darley, W. M., Holm-Hansen, O., and Volcani, B. E. (1967b). *Plant Physiol.*, **42**, 1601.
Coombs, J., Halicki, P. J., Holm-Hansen, O., and Volcani, B. E. (1967c). *Exp. Cell Res.*, **47**, 302.
Coombs, J., Halicki, P. J., Holm-Hansen, O., and Volcani, B. E. (1967d). *Exp. Cell Res.*, **47**, 315.
Coombs, J. and Volcani, B. E. (1968). *Planta.*, **82**, 380.
Corliss, J. B., Dymond, J., Gordon, L. I., Edmond, J. M., von Herzen, R. P., Ballard, R. D., Green, K., Williams, D., Bainbridge, A., and Crane, K. (1979). *Science.*, **203**, 1073.
Craig, H., Chung, Y., and Faideiro, M. (1972). *Earth Planet. Sci. Lett.*, **16**, 50.
Darley, W. M. and Volcani, B. E. (1969). *Experim. Cell. Res.*, **58**, 334.
Davis, E. F. (1918). *Univ. California Publ. Dept. Geol. Sci. Bull.*, **11**, 235–432.
Deacon, G. E. R. (1963). *In* "The Sea", Vol. 2 (Hill, M. N., ed.). Wiley Interscience, New York.
Deffeyes, K. S. (1959). *J. Sedim. Petrol.*, **29**, 602–609.
DeMaster, D. J. (1981). *Geochim. Cosmochim. Acta*, **45**, 1715.
Desikachary, T. V. (1957). *J. Roy. Microscop. Soc.*, **76**, 9.
Donnay, J. D. H. (1936). *Ann. Soc. Geol. Belg.* **59**, B289.
Dugdale, R. C. (1972). *Geoforum*, **11**, 47–61.
Duthie, H. C. and Sreenivassa, M. R. (1971). *Proc. 14th Conf. Great Lakes Res.* 1971, 1.

Duthie, H. C. and Sreenivassa, M. R. (1972). *Proc. 5th Conf. Great Lakes Res.* 1972, 45.
Dwyer, F. P. and Mellor, D. P. (1933). *J. Roy. Soc. New South Wales*, **66**, 378.
Edmond, J. M. (1973). *Nature*, **241**, 391.
Edmond, J. M., Jacobs, S. S., Gordon, A. L., Mantyla, A. W., and Weiss, R. F. (1979a). *J. Geophys. Res.*, **84**, 7809.
Edmond, J. M., Measures, C., McDuff, R. E., Chan, L. H., Collier, R., Grant, B., Gordon, L. I., and Corliss, J. B. (1979b). *Earth Planet. Sci. Letters*, **46**, 1.
Edwards, A. M. C. and Liss, P. S. (1973). *Nature*, **243**, 341.
Eggiman, D. W., Manheim, F. T., and Betzer, P. R. (1979). *Sediment. Petrol.*, **49**, 215.
Einsele, W. and Grim, J. (1938). *Z. Botan.*, **32**, 545.
Eisma, D., and Van der Gaast, S. J. (1971). *Netherlands J. Sea Res.*, **5**, 382.
Ellenburg, J. and Kuhn, G. (1966). *Hall. Jb. f. Mitteldt. Erdg.*, **8**, 67.
Ernst, W. G. and Calvert, S. E. (1969). *Amer. J. Sci.*, **267A**, 114.
Eugster, H. P. (1967). *Science*, **157**, 1177.
Eugster, H. P. (1969). *Contr. Mineral Petrol.*, **22**, 1.
Eugster, H. P. and Jones, B. F. (1968). *Science*, **161**, 160.
Eugster, H. P. and Chou, I. M. (1973). *Econ. Geol.*, **68**, 1144–1168.
Fanning, K. A. and Pilson, M. E. Q. (1971). *Science*, **173**, 1228.
Fanning, K. A. and Pilson, M. E. Q. (1973). *Geochim. Cosmochim. Acta.*, **37**, 2405.
Fanning, K. A. and Pilson, M. E. (1974). *Geophys. Res.*, **79**, 1293.
Fanning, K. A. and Schink, D. R. (1969). *Limnol. and Oceanog.*, **14**, 59.
Flörke, O. W. (1955a). *Neu. J. Mineral. Mh.*, 217.
Flörke, O. W. (1955b). *Ber. Deutsch. Keram, Gessell.*, **32**, 369.
Flörke, O. W., Jones, J. B., and Segnit, E. R. (1975). *Neu Jahrb. Mineral. Mh.*, 369.
Flörke, O. W., Hollmann, R., von Rad, U., and Rösch, H. (1976). *Contrib. Mineral. Petrol.*, **58**, 235.
Folk, R. L. and Weaver, C. E. (1952). *Amer. J. Sci.*, **250**, 498.
Frondel, C. (1962). "The System of Mineralogy", Vol. 3, Silica Minerals. John Wiley, New York.
Gade, M. M., Kirsch, H., and Pollmann, S. (1963). *Neu. Jahrb. Mineral. Mh.*, **100**, 43.
Garavelli, C. L. (1964). *Atti. Della Societa Toscana di Scienze Naturali.*, **71**, 133.
Garrels, R. M. (1965). *Science*, **148**, 69.
Garrels, R. M. and Mackenzie, F. T. (1967). *In* "Equilibrium Concepts in Natural Water Systems" (Gould, R. F., ed.). Amer. Chem. Soc., 222–242.
Garrels, R. M. and Mackenzie, F. T. (1971). "Evolution of Sedimentary Rocks". W. W. Norton, New York.
Garrison, R. E. and Douglas, R. G. (1981). (Editors). "The Monterey Formation and Related Siliceous Rocks of California". Soc. Econ. Paleont. Mineral., Los Angeles.
Gessner, F. (1959). "Hydrobotanik", Vol. 2, Veb. Deutsch. Verlag der Wissensch, Berlin.
Gibson, T. G. and Towe, K. M. *Science*, **172**, 152–154.
Gilbert, J. Y. and Allen, W. E. (1943). *J. Mar. Res.*, **5**, 89.
Goldberg, E. D. (1958). *J. Mar. Res.*, **17**, 178.
Goll, R. M. and Bjørklund, K. R. (1971). *Micropaleontology.*, 17, 434.
Goll, R. M. and Bjørklund, K. R. (1974). *Micropaleontology*, **20**, 38.
Greenwood, R. (1973). *J. Sedim. Petrol.*, **43**, 700.
Gregor, B. (1968). *Nature*, **219**, 360.
Greigg, J. W. (1932). *J. Amer. Chem. Soc..*, **54**, 2846.

Grill, E. V. (1970). *Deep-Sea Res.*, **17**, 245.

Gross, M. G. (1967a). *In* "Estuaries" (Lauff, G. H., ed.). Amer. Ass. for Adv. of Sc. **273**, New York.

Gross, M. G. (1967b). *Nature*, **216**, 670.

Grunau, H. R. (1965). *Eclogae Geol. Helvet.*, **58**, 157.

Gruner, J. W. (1940). *Econ. Geol.*, **35**, 867.

Guinasso, N. L. and Schink, D. R. (1975). *J. Geophys. Res.* **80**, 3032.

Harper, H. E. and Knoll, A. H. (1975). *Geology*, **3**, 175.

Harriss, R. C. (1966). *Nature*, **212**, 275.

Hay, R. L. (1966). *Geol. Soc. Amer. Spec. Paper.*, **85**, 130 pp.

Hay, R. L. (1968). *Contr. Mineral. and Petrol.*, **17**, 255.

Healey, F. P., Coombs, J., and Volcani, B. E. (1967). *Ark. F. Mikrobiol.*, **59**, 131.

Heath, G. R. (1968). Ph.D. Dissertation, Univ. of California, San Diego. 168 pp.

Heath, G. R. (1973). *In* "Initial Reports of the Deep Sea Drilling Project", (Van Andel *et al.*,) **16**, US Govt. Printing Office, Washington DC.

Heath, G. R. (1974). *In* "Studies in Paleo-Oceanography" (Hay, W. W., ed.). *Soc. Econ. Paleont. Mineral. Spec. Publ.* **20**, 77.

Heath, G. R. and Dymond, J. (1973). *Geology*, **3**, 181.

Heath, G. R. and Moberley, R. (1971). *In* "Initial Reports of the Deep Sea Drilling Project" (Winterer, E. L. *et al.*,) **7**, Part. 2. US Government Printing Office, Washington DC.

Heath, G., Moore, T. C. Jr., and Dauphin, J. Paul. (1976). *Geol. Soc. Amer. Memoir*, **145**, 393.

Hein, J. R., Scholl, D. W., Barron, J. A., Jones, M. G., and Miller, J. (1978). *Sedimentology*, **25**, 155.

Helgeson, H. C. and Mackenzie, F. T. (1970). *Deep-Sea Res.*, **17**, 877–892.

Heron, S. D., Robinson, G. C. and Johnson, H. S. (1965). *State Devel. Board South Carolina, Divn. Geol. Bull.*, **31**, 64 pp.

Hewett, D. F. (1917). *J. Wash. Acad. Sci.* **7**, 196.

Heydemann, A. (1964). *Beitr. zur Mineral. und Petrog.*, **10**, 242.

Honjo, S. (1980). *J. Marine Res.*, **38**, 53–97.

Hurd, D. C. (1972). *Hawaii Inst. Geophys. Publ.*, 72-22, 81 pp.

Hurd, D. C. (1973). *Geochim. Cosmochim. Acta.*, **37**, 2257.

Hurd, D. C. (1977). *Geochim. Cosmochim. Acta.*, **41**, 1213.

Hurd, D. C. and Theyer, F. (1975). *In* "Advances in Chemistry Series No. 147" (Gibb, R. P., Jr., ed.). American Chemical Society, 211.

Hurd, D. C. and Theyer, F. (1977). *Amer. J. Sci.* **277**, 1168.

Hutchinson, G. E. (1967). "A Treatise on Limnology". Vol. 2. John Wiley, New York.

Iijima, A. and Utada, M. (1966). *Sedimentology*, **7**, 327–357.

Iler, R. K. (1955). "The Colloid Chemistry of Silica and Silicates". Cornell Univ. Press.

Inoue, M. (1973). *J. Geol. Soc. Japan*, **79**, 277.

Johnson, D. A. and Nigrini, C. (1980). *Marine Micropaleo.* **5**, 111–152.

Johnson, T. C. and Eisenreich, S. J. (1979). *Geochim. Cosmochim. Acta.*, **43**, 77.

Joides, (1967). *Bull. Amer. Assoc. Petrol. Geol.*, **51**, 1787.

Jones, J. B. and Segnit, E. R. (1971). *J. Geol. Soc. Australia.*, **18**, 57.

Jones, J. B. and Segnit, E. R. (1972). *J. Geol. Soc. Australia.*, **18**, 419.

Jones, J. B. and Segnit, E. R. (1975). *Contrib. Mineral. Petrol.*, **51**, 231.

Jones, J. B., Sanders, J. V., and Segnit, E. R. (1964). *Nature*, **204**, 990.

Jones, L. H. P. and Milne, A. A. (1963). *Plant and Soil*, **17**, 207.
Jørgensen, E. G. (1955). *Physiol. Plantarum*, **8**, 840.
Kamatani, A. (1971). *Marine Biol.*, **8**, 89.
Kamatani, A. (1974). *J. Oceanogr. Soc. Japan.*, **30**, 147.
Kanaya, T. and Koisumi, I. (1966). *Science Reports of Tokohu University, Sendai.* 2nd Series (Geology). **37**, 89.
Kastner, M. and Keene, J. R. (1975). *Proc. IXth Int. Cong. of Sedimentology*, **7**, 89.
Kastner, M., Keene, J. B., and Gieskes, J. M. (1977). *Geochim. Cosmochim. Acta.*, **41**, 1041.
Kato, K. (1969). *Geochem. J.*, **3**, 87.
Keene, J. B. (1976). PhD Thesis, Univ. of California, 264 pp.
Kennedy, V. C. (1971). *In* "Nonequilibrium Systems in Natural Water Chemistry". Advances in Chemistry Series **106**, 94–130.
Kido, K. (1974). *Mar. Chem.*, **2**, 277.
Kido, K. and Nishimura, M. (1973). *J. Oceanogr. Soc. Japan.*, **29**, 185.
Klasik, J. A. (1975). *Science*, **189**, 631.
Knauss, J. A. (1962). *J. Geophys. Res.*, **67**, 3943.
Kolbe, R. W. (1955). *Repts. Swedish Deep-Sea Exped.*, **7**, 151.
Kozlova, O. G. and Mukhina, V. V. (1967). *Int. Geol. Rev.*, **9**, 1322.
Krauskopf, K. B. (1956). *Geochim. Cosmochim. Acta.*, **10**, 1.
Lal, D. and Lerman, A. (1973). *J. Geophys. Res.* **78**, 7100.
Lancelot, Y. (1973). *In* "Initial Reports Deep Sea Drilling Project". Vol. XVII (Winterer, E. L., Ewing, J. E. *et al.*). US Govt. Printing Office, Washington DC.
Lanning. F. C., Ponnaiya, B. W. X., and Crumpton, C. F. (1958). *Plant. Physiol.*, **33**, 339.
Laurent, E. and Scheere, J. (1971). *Bull. Belg. Ver. Geol. Paleont. Hydrol.*, **80**, 145.
Leinen, M. (1977). *Geochim. Cosmochim. Acta.*, **41**, 671.
Lerman, A. and Mackenzie, F. T. (1975). *Earth & Planet Sci.*, **25**, 82–88.
Lerman, A. and Lal, D. (1977). *Am. J. Sci.*, **277**, 238.
Levin, J. and Ott, E. (1933). *Z. Krist.*, **85**, 305.
Levitan, M. A., Dontsova, E. I., Lisitsin, A. P., and Bogdanov, Y. A. (1975). *Geochem. Int.*, **12**, 95.
Lewin, J. C. (1954). *J. Gen. Physiol.*, **37**, 589.
Lewin, J. C. (1955a). *Plant Physiol.*, **30**, 129.
Lewin, J. C. (1955b). *J. Gen. Physiol.*, **39**, 1.
Lewin, J. C. (1957). *Can. J. Microbiol.*, **3**, 427.
Lewin, J. C. (1961). *Geochim. Cosmochim. Acta.*, **21**, 182.
Lewin, J. C. (1962). *In* "Physiology and Biochemistry of Algae" (Lewin, R. A., ed.). Academic Press, London and New York.
Lewin, J. C. (1966). *Phycologia*, **6**, 1.
Lisitsin, A. P. (1967). *Int. Geol. Rev.*, **9**, 631, 842, 980, 1114.
Lisitsin, A. P. (1971). *In* "The Micropaleontology of Oceans" (Funnell, B. M. and Riedel, W. R., eds.). Cambridge Univ. Press, Cambridge.
Lisitsin, A. P., Belyayev, Y. I., Bogdanov, Y. A., and Bogoyavlenskiy, A. N. (1967). *Intern. Geol. Rev.*, **9**, 604.
Liss, P. S. and Spencer, C. P. (1970). *Geochim. Cosmochim. Acta.* **35**, 1073.
Livingston, D. A. (1963). *U.S. Geol. Survey Prof. Paper.* 440-G. 64 pp.
Lund, J. W. G. (1964). *Verh. Internat. Verein. Limnol.*, **XV**, 37.
McAtee, J. L., House, R., and Engster, H. P. (1968). *Am. Mineralogist.*, **53**, 2061.
McCave, I. N. (1975). *Deep-Sea Res.*, **22**, 491.

Mackenzie, F. T. and Garrels, R. M. (1965). *Science*, **150**, 57.
Mackenzie, F. T. and Garrels, R. M. (1966a). *Amer. J. Sci.*, **264**, 507.
Mackenzie, F. T. and Garrels, R. M. (1966b). *J. Sedim. Petrol.*, **36**, 1075.
Mackenzie, F. T., Garrels, R. M., Bricker, O. P., and Bickley, F. (1967). *Science*, **155**, 1404.
Mackenzie, F. T. and Gees, R. (1971). *Science*, **173**, 533.
Mallard, M. E. (1890). *Bull. Soc. Fr. Mineral. Cristallog.*, **13**, 63.
Martin, J. H. and Knauer, G. A. (1973). *Geochim. Cosmochim. Acta.*, **37**, 1639.
Mattson, P. H. and Pessagno, E. A. (1971). *Science*, **174**, 138–139.
Mayer, L. M. and Gloss, S. P. (1980). *Limnology and Oceanogr.*, **25**, 12.
Maynard, J. B. (1976). *Geochim. Cosmochim. Acta.*, **40**, 1523–1532.
Maynard, N. (1976). *Paleobiology*, **2**, 99.
Meybeck, M. (1979). *Rev. Géol. Dynam. Géog. Phys.*, **21**, 215.
Miller, R. W. (1967). *Soil Sci. Soc. Amer. Proc.*, **31**, 46.
Milliman, J. D. and Boyle, E. (1975). *Science*, **189**, 995.
Mitsui, K. and Taguchi, K. (1977). *J. Sedim. Petrol.*, **47**, 158.
Mizutani, S. (1966). *Nagoya Univ. J. Earth Sci.*, **14**, 56.
Mizutani, S. (1977). *Contrib. Min. Petrol.*, **61**, 129.
Moore, T. C. (1978). *Mar. Micropaleontol.*, **3**, 229.
Moore, T. C., Jr., Heath, G. R., and Kowsmann, R. O. (1973). *J. Geol.*, **81**, 458.
Mortimer, C. H. (1941). *J. Ecol.*, **29**, 280.
Mortimer, C. H. (1942). *J. Ecol.*, **30**, 147.
Murata, K. J. (1940). *Amer. J. Sci.*, **238**, 586–596.
Murata, K. J. and Nakata, J. K. (1974). *Science*, **184**, 567.
Murata, K. J. and Larson, R. R. (1975). *J. Res. U.S. Geol. Survey*, **3**, 553.
Murata, K. J. and Randall, R. G. (1975). *J. Res. U.S. Geol. Survey*, **3**, 567.
Murata, K. J., Friedman, I., and Gleason, J. D. (1977). *Am. J. Sci.*, **277**, 259.
Murata, K. J., Dibblee, T. W., and Drinkwater, J. L. (1979). *U.S. Geol. Survey. Prof. Paper* 1082, 8 pp.
Murray, J. and Renard, A. F. (1891). Deep Sea Deposits, Neill, Edinburgh.
Nriagu, J. O. (1978). *Limnol. Oceanogr.*, **23**, 53.
Oehler, J. H. (1973). *Nature Phys. Science.*, 241, 64.
Oehler, J. H. (1975). *J. Sedim. Petrol.*, **45**, 252.
Omure, M., Ookawara, S., and Iwai, S. H. (1956). *Naturwiss.*, **43**, 495.
O'Neil, J. R. and Hay, R. L. (1973). *Earth & Plan. Sci. Letters.*, **19**, 257.
Parker, J. I. and Edgington, D. N. (1976). *Limnol. Oceanogr.*, **21**, 887.
Parker, J. I., Conway, H. L., and Yaguchi, E. M. (1977). *J. Fish. Res. Board Canada.*, **34**, 545.
Perry, E. A. and Hower, J. (1970). *Clays and Clay Min.*, **18**, 165–177.
Perry, E. A. (1971). *Deep-Sea Res.*, **18**, 921–924.
Peterson, D. H., Conomos, T., Broenkow, W., and Scrivani, E. (1975). *In* "Estuarine Research", Vol. 1 (Cronin, L. E., ed.). Academic Press, London and New York.
Petrushevskaya, M. G. (1971). *In* "Micropaleontology of Oceans" (Funnell, B. M. and Riedel, W. R., eds.). Cambridge Univ. Press, Cambridge.
Pettijohn, F. J. (1957). "Sedimentary Rocks", Harper N.Y., New York.
Platt, T. and Subba Rao, D. V. (1975). *In* "Photosynthesis and Productivity in Different Environments", *Intern. Biol. Prog.*, Cambridge University Press, Cambridge.
Pratje, O. (1951). *Deutsch. Hydrog. Z.*, **4**, 1.
Pytkowicz, R. M. (1967). *Geochim. Cosmochim. Acta.*, **31**, 63–73.

Pytkowicz, R. M. (1975). *Earth Sci. Rev.*, **11**, 1–46.
Redfield, A. C. (1958). *Amer. Sci.*, **46**, 205–221.
Redfield, A. C., Ketchum, B. H., and Richards, F. A. (1963). *In* "The Sea", Vol. 2 (Hill, M. N., ed.). Interscience, New York.
Reimann, B. E. F. (1964). *Experim. Cell Res.*, **34**, 605.
Reimann, B. E. F., Lewin, J. C., and Volcani, B. E. (1965). *J. Cell Biol.*, **24**, 39.
Reimann, B. E. F., Lewin, J. C., and Volcani, B. E. (1966). *J. Phycol.*, **2**, 74.
Renz, G. W. (1976). *Bull. Scripps Inst. Oceanogr.*, **22**, 267 pp.
Reynolds, R. C. and Anderson, D. M. *J. Sedim. Petrol.*, **37**, 966–969.
Reynolds, W. R. (1970). *J. Sedim. Petrol.*, **40**, 829–838.
Riedel, W. R. (1959). *In* Special Publ. 7, Soc. Econ. Paleontol. Mineral., 80.
Riech, V. and von Rad, U. (1979). *In* "Deep Drilling Results in the Atlantic Ocean: Continental Margins and Paleoenvironment" (Talwani, M., Hay, W. W., and Ryan, W. B. F., eds.). Amer. Geophys. Union, Washington DC.
Robertson, A. H. F. (1977). *Sedimentology* **24**, 11.
Rogall, E. (1939). *Planta*, **29**, 279.
Round, F. E. (1968). *J. Exp. Marine Biol. Ecol.*, **2**, 64.
Russell, K. L. (1970). *Geochim. Cosmochim. Acta.*, **34**, 893–907.
Scheere, J. and Laurent, E. (1970). *Bull. Belg. Ver. Geol. Paleont. Hydrol.*, **70**, 225.
Schink, D. R. (1967). *Geochim. Cosmochim. Acta.*, **31**, 897.
Schink, D. R. and Guinasso, N. L. (1977). *Marine Geol.*, **23**, 133.
Schink, D. R., Fanning, K. A., and Pilson, M. E. Q. (1974). *J. Geophys. Res.*, **79**, 2243.
Schink, D. R., Guinasso, N. L., and Fanning, K. A. (1975). *J. Geophys. Res.*, **80**, 3013.
Scholl, D. W. and Creager, J. S. (1973). *In* "Initial Repts. Deep Sea Drilling Project", Vol. XIX (Creager, J. S. and Scholl, D. W., eds.). 897–913 US Government Printing Office, Washington DC.
Schrader, H. J. (1971). *Science*, **174**, 55.
Schutz, D. F. and Turekian, K. K. (1965). *Geochim. Cosmochim. Acta.*, **29**, 259.
Sholkovitz, E. (1973). *Geochim. Cosmochim. Acta.*, **37**, 2043.
Sieffert, B. and Wey, R. (1967). *Silicates Ind.*, **32**, 415.
Siever, R. (1957). *Am. Mineral.*, **42**, 821.
Siever, R. (1968a). *Sedimentol.*, **11**, 5–29.
Siever, R. (1968b). *Earth Planet. Sci. Letters.*, **5**, 106.
Siever, R., Beck, K. C., and Berner, R. A. (1965). *J. Geol.*, **73**, 39.
Siever, R. and Woodford, N. (1973). *Geochim. Cosmochim. Acta.* **37**.
Sillen, L. G. (1961). *In* "Oceanography" (Sears, M., ed.). Amer. Assoc. Adv. Sci. Publ. **67**, 549.
Smith, W. E. (1960). *Geol. Mijnbouw.*, **39**, 1.
Sosman, R. B. (1954). *Science*, **119**, 738.
Stein, C. L. and Kirkpatrick, R. J. (1976). *J. Sedim. Petrol.*, **46**, 430.
Sterling, C. (1967). *Am. J. Botany*, **54**, 840.
Stishov, S. M. and Popova, S. V. (1961). *Geokhimiya*, **10**, 837.
Stoermer, E. F. and Pankratz, H. S. (1964). *Am. J. Botany*, **51**, 986.
Stoermer, E. F., Pankratz, H. S., and Drum, R. W. (1964). *Protoplasma*, **59**, 1.
Stoermer, E. F., Pankratz, H. S., and Bowen, C. C. (1965). *Amer. J. Botany*, **52**, 1067.
Stommel, H. and Arons, A. B. (1960). *Deep Sea Res.*, **6**, 217.
Sverdrup, H. U. (1938). *J. Mar. Res.*, **1**, 155.

Swineford, A. and Franks, P. C. (1959). *In* Special Publ. 7, Soc. Econ. Paleont. Mineral., 111–120.
Takahashi, K. and Honjo, S. (1981). *Micropaleontology*, **27**, 140–190.
Taliaferro, N. L. (1933). *Univ. California Publ. Dept. Geol. Sci. Bull.*, **23**, 1–55.
Taliaferro, N. L. (1934). *Bull. Geol. Soc. Amer.*, **45**, 189.
Tessenow, U. (1966). *Arch. Hydrobiol.*/Suppl. XXXII, 1.
Turekian, K. K. (1971). *In* "Impingement of Man on the Oceans" (Hood, D. W., ed.). John Wiley, New York.
Turpin, D. H. and Harrison, P. J. (1979). *J. Exp. Mar. Biol. Ecol.*, **39**, 151.
Vinogradov, A. P. (1953). Sears Found. Mar. Res. Memoir II. Yale Univ. Press, New Haven.
Von Rad, U. and Rösch, H. (1972). *In* "Initial Repts. Deep Sea Drilling Project", Vol. XIV (Hayes, D. E. *et al.*). U.S. Govt. Printing Office, 727, Washington DC.
Von Rad, U. and Rösch, H. (1974). *In* "Pelagic Sediments: on land and under the sea" (Hsii, K. J. and Jenkyns, H. C., eds.). *Spec. Publs. Int. Ass. Sediment.*, **1**, 327.
Von Rad, U., Riech, V. and Rösch, H. (1977). *In* "Initial Report of the Deep Sea Drilling Project" (Y. Lancelot and E. Seibold, eds.). **41**, 879.
Warren, B. E. and Biscoe, J. (1938). *J. Amer. Ceram. Soc.*, **21**, 49.
Weaver, F. M. and Wise, S. W. (1972). *Nature Physical Science*, **237**, 56.
Weaver, F. M. and Wise, S. W. (1974). *Science*, **184**, 899.
Weiler, R. R. (1973). *Limnol. Oceanog.*, **18**, 918.
Werner, D. (1977). "The Biology of Diatoms", Univ. Calif. Press, California.
Wetzel, R. G. (1975). "Limnology", W. B. Saunders, Toronto.
Wilson, M. J., Russell, J. D., and Tait, J. M. (1974). *Contrib. Mineral. Petrol.*, **47**, 1.
Wise, S. W. and Kelts, K. R. (1972). *Trans. Gulf Coast Assoc. Geol. Soc.*, **22**, 177.
Wise, S. W., Buie, B. F., and Weaver, F. M. (1972). *Ecol. Geol. Helv.*, **65**, 157.
Wise, S. W. and Weaver, F. M. (1973). *Trans. Gulf Coast Assocn. Geol. Soc.*, **23**, 305.
Wise, S. W. and Weaver, F. M. (1974). *In* "Pelagic Sediments: on land and under the sea" (Hsü, K. J. and Jenkyns, H. C., eds.). *Spec. Publ. Int. Assoc. Sedimentol.*, **1**, 301.
Wollast, R. *In* "The Sea", Vol. 5 (Goldberg, E. D., ed.). J. Wiley, New York.
Wollast, R. and DeBroeu. (1971). *Geochim. Cosmochim. Acta.*, **35**, 613.
Wong, G. T. F. and Grosch, C. E. (1978). *J. Marine Res.*, **36**, 735.
Yemel 'Yanov, Y. M. (1975). *Doklady Akad. Nauk. S.S.S.R. Earth Sci. Sect.*, **225**, 227.

6

Physical and Chemical Properties of Siliceous Skeletons

D. C. Hurd

Shell Development Company
Geological Engineering Section
Houston, USA

6.1   INTRODUCTION

Having read some five chapters about the importance of silica in a variety of environments, the reader may now wonder why a separate chapter should be devoted to the chemical and physical properties of siliceous skeletons. The answer lies in the general agreement that organisms quite regularly perform precipitation processes of considerable magnitude that would never occur spontaneously under the conditions in which the organisms are operating. The oceans, rivers, and lakes are all quite undersaturated with respect to the solubility of biogenic silica, yet diatoms, radiolarians, silicoflagellates, and sponges together probably precipitate on the order of 10^{16} g of silica annually. Most of this material seems to dissolve and does not enter into the sedimentary record. There are many ways of describing the rates of dissolution and the various physical properties of biogenic silica depending on the purpose of the investigator. What follows is a series of discussions about the siliceous skeletons of mostly radiolarians and diatoms as applied to understanding the solid itself and in formulating re-cycling models which are consistent with the physical reality of the solid. At the end of some sections, I have also tried to pose additional questions regarding future research about biogenic silica using various techniques and, where suitable, references which may offer a useful beginning for the researcher have been included.

6.2 SOLUBILITY OF BIOGENICALLY PRECIPITATED SILICA

A number of authors (Krauskopf, 1956; Krauskopf, 1959; Okamoto, Okura, and Goto, 1957; Siever, 1962; Mackenzie and Gees, 1971) have studied the solubility of biogenically precipitated silica. The solubility of artificially precipitated silica (silica gel) and of vitreous silica, two forms of silica which have similar solubilities, have also been studied (Hurd, 1973; Greenberg, 1957; Morey, Fournier, and Rowe, 1964; Fournier and Rowe, 1962; Morey, Fournier, and Rowe, 1962; Fournier, 1973; Harder, 1971; Grill and Richards, 1964; Kamatani, 1971; Baumann, 1965; Wollast, 1974). Biogenic

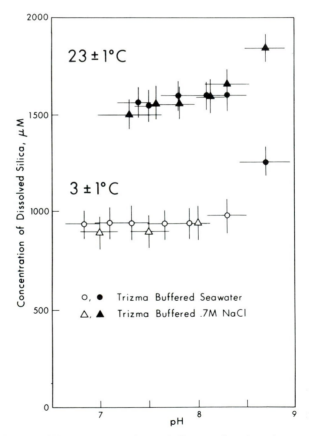

FIG. 6.1   Solubility of biogenically precipitated silica as a function of temperature and pH in both buffered seawater and sodium chloride solution (from Hurd, 1973).

silica and silica gel probably resemble each other more than either resembles vitreous silica; excellent discussions of silica gel and vitreous silica properties may be found in Sosman (1965) and Iler (1979). Depending on the manner of preparation, degree of internal ordering, and specific surface area, the solubilities of vitreous silica and silica gel vary widely; however, both substances share similar ranges which are at least one order of magnitude greater than quartz in the 0–25°C range of temperatures considered here. Equilibrium values for biogenic silica in seawater, pH 7.5–8.3, are in the range 1500–1700 µM at 23° ± 1°C, and 800–1000 µM at 3 ± 1°C as shown in Fig. 6.1 (Hurd, 1973).

Additional considerations regarding the change in solubility with age of biogenic silica and of its solubility relationships relative to other silica polymorphs can be found in Hurd and Theyer (1975), Walther and Helgeson (1977), and references therein.

The effect of pressure on the solubility of biogenic silica and silica gel has been studied by Jones and Pytcowicz (1973), Willey (1974), and Griffin (1980), and summarized by Walther and Helgeson (1977) (although the latter authors appear to have mis-plotted Hurd's 1973 data in their Fig. 3 compared with the above data in Fig. 6.1). Figure 6.2 shows the effect of pressure on the solubility of biogenic silica and silica gel (Griffin, 1980) at 3 ± 1°C. In Fig. 6.3 a semi-empirical approach has been taken toward estimating the combined effects of temperature and pressure with depth in the

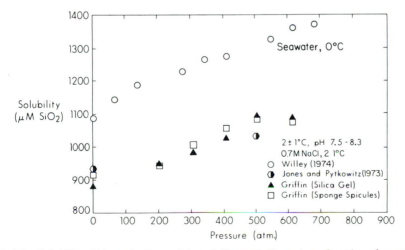

FIG. 6.2   Solubility of biogenically precipitated silica and silica gel as a function of pressure at about 2°C (from Griffin, 1980).

water column. A section through the thermocline has been taken as an example and it is assumed that the effects of temperature and pressure are independent of one another. Solubility as a function of temperature has been estimated using the Gibbs free energy equation and a value of 3820 cal mol^{-1} for biogenic silica (Hurd and Theyer, 1975). The effect of pressure relative to one atmosphere is approximately estimated from Fig. 6.2 using the ratio (Solubility$_P$/Solubility$_{1\ atm.}$) multiplied by the estimated solubility at a given temperature. It can be seen that the solubility undergoes a minimum at about 1500 m and then slowly increases. At high latitudes where water column temperatures are often nearly constant, the solubility would show a linear increase with depth. In general in the range of 0–25°C, 1–600 atm., the effects of temperature are three to four times larger than those for pressure. In the pH range 5–8.3, the solubility of biogenic silica parallels that of silica gel and is essentially constant.

The adsorption of cations on the surface of biogenic silica can greatly alter its solubility as demonstrated by Lewin (1961) and calculated by Hurd

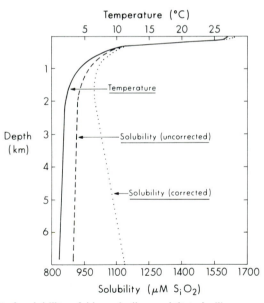

FIG. 6.3 Estimated solubility of biogenically precipitated silica as a function of both temperature and depth in the Central Equatorial Pacific. Solid line = distribution of temperature with depth; dashed line = solubility estimated only for changes in temperature; and dotted line = solubility estimated for both change in temperature and pressure (from Griffin, 1980).

(1973). Many investigators have adsorbed a wide variety of cations onto silica gel but so far few, if any, have done so using the *actual* seawater concentrations of these cations. Whether or not cations actually adsorb onto silica at their seawater concentrations and, if they do so, their rate of adsorption are both quite important topics because of the variety of time scales involved in the re-cycling of silica, for example, in the oceans.

The actual skeletons may be formed within a few hours to weeks, require days to months to settle through the water column, be in the nepheloid layer and at the sediment–water interface for months to several hundreds of years, and in the upper 10–15 cm of the sediment for thousands to hundreds of thousands of years, where they can still dissolve and the silica be removed by diffusion. Also, in each successive case the proximity of biogenic silica and alumino-silicate particles increases until they are finally in constant contact in the sediment. Almost no cores whose pore waters have been analysed *in situ* or at least at bottom water temperatures have been found to be saturated with respect to biogenic silica, even though substantial amounts of siliceous skeletons may have been present, a possible exception being the nearly pure diatom oozes from the Antarctic analysed by DeMaster (1979). In any case, adsorption of some sort occurs but it is still unclear when the adsorption occurs, how much is adsorbed, and over what time scale this process takes place.

6.2.1 Questions Relative to the Solubility Section

1. What is silica solubility as a function of percent surface coverage by various cations? Iler (1979); Brewer and Hao (1979).
2. Can the solubility of amorphous silica be affected by adsorption of cations at concentration levels found in the oceans? Iler (1979, pp. 79–82), Willey (1975a,b).
3. Assuming that there are a variety of cations naturally associated with the silica frustule, (a) are these elements uniformly distributed throughout the silica matrix or patchy or in layers; (b) as the solid dissolves, do these cations tend to remain behind forming a solubility/solution rate inhibiting layer or are they removed congruently with the silica?
4. Do the various particle and pore size distributions which naturally exist in biogenic silica have a measurable effect on its solubility? If so, do these distributions change with time *within* the skeleton and are there attendant solubility changes which also occur?
5. Do the quite small amounts of organic matter naturally occurring in siliceous skeletons have an effect on the solubility/dissolution rate? What is the method and significance (if any) of its incorporation? King (1975), Hecky *et al.* (1973).

6.2.2 Factors Affecting Dissolution Rates

The first order congruent solution of amorphous silica is described by the equation:

$$dC_{sol}/dt = k_2(C_{sat} - C_{sol})S \tag{1}$$

where k_2 is the first-order rate constant in cm s^{-1}; C_{sat} and C_{sol} are the concentrations of a solution saturated at a particular temperature and pH, and the solution being observed at time t in mol cm^{-3}; and S is the surface area of the solid per unit volume of solution in cm^{-1} (O'Connor and Greenberg, 1958; van Lier, de Bruyn, and Overbeek, 1960; Hurd, 1972; Wollast, 1974). This equation is identical to the type used if diffusion control of solution is assumed; Wollast and van Lier argue from a theoretical standpoint that diffusion control of solution need *not* follow from such an equation. Furthermore, the following laboratory evidence suggests that chemical mechanisms involving the release of the silica molecule from the amorphous silica matrix have a greater effect on the solution rate than the rate at which the molecule diffuses away from the surface through the solvent: a solid undergoing first-order diffusion-controlled solution obeys the following equation:

$$\frac{dC_{sol}}{dt} = \frac{D}{\delta}(C_{sat} - C_{sol})\frac{A}{V} \tag{2}$$

where D is the diffusion coefficient of the diffusing species in cm^2 s^{-1}, δ is the

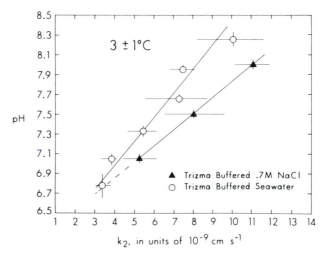

FIG. 6.4   Dissolution rate constant, k_2, as a function of pH in both sea water and sodium chloride solutions at about 3°C (figure redrafted from Hurd, 1973).

thickness of the Nernst layer in cm, A the area of the solid in cm^2 and V the volume of the solution, cm^3 (van Lier, de Bruyn and Overbeek, 1960). The similarity to (1) is obvious: $S = A/V$ and if the process is diffusion controlled $k_2 = D/\delta$. Figures 6.1, 6.4, and 6.5 give the variation of C_{sat} and k_2 in equation (1) as a function of temperature and pH for biogenic silica. The rate constant can be seen to vary nearly two orders of magnitude from pH 8.3, 23°C (surface seawater, tropical-subtropical areas) to pH 6.9–7.3 2–3°C (pore waters of deep-sea sediments). The saturation values are essentially constant within the above pH range at a given temperature but vary by about a factor of two from 0°C to 25°C. By using an upper limit for k_2 of 10^{-7} cm s^{-1} (pH 8.3, 23°C), and 10^{-5} cm^2 s^{-1} for the diffusion coefficient of dissolved silica in sea water at 25°C (Wollast and Garrels, 1971) a value is arrived at for the thickness of the Nernst layer δ of 10^2 cm for biogenic silica. Since for small particles falling through solution the Nernst layer thickness is on the order of the size of the particle (Levich, 1962) and the radiolarians in this study were about 10^{-2} cm in diameter, one must conclude (from a diffusion-controlled viewpoint) that only a small portion of the surface is geometrically available for solution—from the above calculations only 0.01%. However, as Wollast (1974) has shown, the value for k_2 (25°, pH 8–8.3) is only twice that found by Stöber (1967) for finely ground amorphous silica under the same conditions suggesting that, geometrically, perhaps only 50% of the BET measured surface area is readily available for solution rather than 0.01% (assuming that the value of k_2 for vitreous silica and biogenic

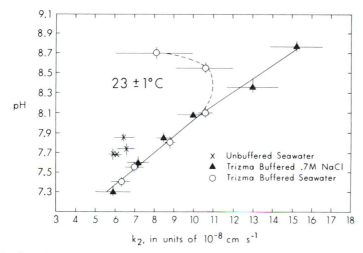

FIG. 6.5   Dissolution rate constant, k_2, as a function of pH in both sea water and sodium chloride solutions at about 23°C (figure redrafted from Hurd, 1973).

silica under identical chemical and geometrical conditions would be the same; this is quite probably *not* true, i.e. the value of k_2 for biogenic silica is probably higher). Similar experiments involving the dissolution rate of vitreous silica as a function of pH and ionic strength have been done by Wirth and Gieskes (1979). Re-calculation of their data to estimate values of k_2 (Fig. 6.6) and inserting their value of k_2 at pH 8, ionic strength 0.7, 25°C into calculations of the type done above for biogenic silica, also agree with

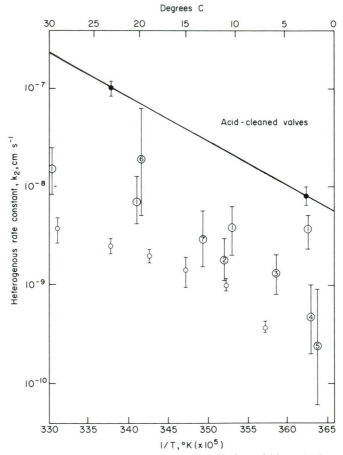

FIG. 6.6   Dissolution rate constants, k_2, for both acid-cleaned biogenically precipitated silica and naturally decomposing phytoplankton as a function of the reciprocal of absolute temperature. Encircled numbers refer to the following investigators: 1, Kamatani, 1969; 2, Grill and Richards, 1964; 3, Lewin, 1961; 4, Kido and Nishimura, 1973, 1975; 5, Edmond, 1974; Matsumoto, 1973; Tsunogai, 1972; 6, Nelson *et al.*, 1973. Unfilled circles are constants calculated from Lawson, Hurd, and Pankratz, 1978; filled circles are from Hurd (1973) (figure from Lawson, Hurd, and Pankratz, 1978).

those of Wollast (1974). Thus it appears that the dissolution rate of biogenic silica as well as other silica polymorphs is controlled by chemical reactions at the solid–liquid interface and not by diffusion processes.

It should be noted that the gross geometric configurations of the tests of silica-secreting organisms in general, and radiolarians and diatoms in particular, can be quite complicated (see for example, Kling, 1978; Burkle, 1978). Our data (to be discussed at greater length in Section 6.5 on surface area and porosity) using transmission electron microscopy (TEM) and gas adsorption suggest that the *inter-connected elements* of siliceous skeletons may also be quite porous depending on the phylum, order, stage of formation or dissolution etc., a fact which has also been implied or described by earlier workers (Lewin, 1967; Weiler and Mills, 1965; Hurd and Theyer, 1975, 1977; Lawson et al., 1978; Hurd et al., 1979; Simpson and Vacarro, 1974; Stoermer et al., 1965; Dawson, 1973; Reimann et al., 1966). So we must wonder whether, during the dissolution process and when bulk solution values are far from saturation with respect to biogenic silica, some of the innermost pores might become saturated with respect to amorphous silica. That is, the rate at which dissolved silica *in the pores* was removed by diffusional processes might be slower than the rate at which dissolved silica was being supplied to the pores by the dissolution of the walls of the pores.

In order to determine this we will need to define several equations generally following the conventions of Frank-Kamenetskii (1969, pp. 98–110).

The basic assumptions are:

1. The thickness of the porous skeletal layer is large enough to be regarded as infinite. The actual case which Frank-Kamenetskii is considering is one where an outer solution contains a dissolved material which is diffusing into and reacting within the porous solid; we are considering somewhat the converse, i.e. the walls of the pores are giving off material which diffuses out through the porous solid into the solution outside. Mathematically these are identical. We are also assuming that the surface area of the solid does not change with time.
2. The reaction of dissolution is first order. This seems to be true for biogenic silica according to Hurd (1972) and our earlier discussion above.
3. Diffusion is considered only in the direction perpendicular to the external solid–solution interface and is essentially planar.
4. After a given lapse of time a stationary state is reached where "a concentration distribution is established within the (porous) material (which does not change)."

For a first-order reaction, the flux across the boundary between the porous solid and the bulk solution is:

$$dm/dt = [(D')(k_2S)]^{1/2}(C_{sat} - C_{sol}) \tag{3}$$

where dm/dt is the flux across the porous solid-bulk solution boundary in $mol\,cm^{-2}\,s^{-1}$, D' is the apparent diffusion coefficient of the solute wending its way through the pores, corrected for adsorption and tortuosity in $cm^2\,s^{-1}$ and the other terms are as defined earlier.

We shall use the following values for the above: $C_{sat} = 1.6 \times 10^{-6}\,mol\,cm^{-3}$, $C_{sol} = 0$, $k_2 \sim 10^{-7}\,cm\,s^{-1}$ ($25°C$, pH 8, $35^0/_{00}$), $S \sim 3 \times 10^6\,cm^{-2}\,cm^{-3}$, so $k_2 S \sim 0.3\,s^{-1}$, $D' \sim 5 \times 10^{-6}\,cm^{-2}\,s^{-1}$. Therefore the maximum flux is ca. $2 \times 10^{-9}\,mol\,cm^{-2}\,s^{-1}$ across the porous solid-bulk solution boundary, but is equal to about $1.6 \times 10^{-13}\,mol\,cm^{-2}\,s^{-1}$ from the walls of the pores (i.e. $k_2 C_{sat} = mol\,cm^{-2}\,s$ as flux from the solid surface).

The depth of penetration of the reaction, L, is approximated by

$$L \sim (D'/k_2 S)^{1/2} \sim 4 \times 10^{-3}\,cm \sim 40\,\mu \qquad (4)$$

and is the approximate depth wherein the total dissolution flux from the walls of the pores within the porous solid overwhelms the diffusive removal processes within the pores and is also the minimum depth necessary to achieve the maximum flux across the porous solid-bulk solution boundary for the system we are considering.

The reader may wonder why H, the actual thickness of the porous layer, even begins to approach L when the actual dissolution rates are so far from the diffusion control limiting dissolution conditions as mentioned earlier. My understanding of Frank-Kamenetskii's arguments is that the flux across the porous solid-bulk solution (PS–BS) interface relative to those reactions occurring within the porous solid (whose magnitude vary with the depth of the porous layer) determine the extent to which all of the porous layer is active or not. As the depth of the porous layer increases, so must the absolute flux across the PS–BS interface until a point is reached where either the turbulent diffusion processes needed to carry away the dissolution products become limiting or, as mentioned above, the total dissolution flux from the walls overwhelms the diffusive removal processes within the pores. At this point also the magnitude of such turbulent diffusion processes will directly affect the concentration gradient from the PS–BS interface into solution by changing the thickness of the relatively quiescent layer near the PS–BS interface (Levich, 1962). For small particles the thickness of such a layer is on the same order as the particle itself (Levich, 1962).

In that the skeletal elements of radiolarians and diatoms are usually not thicker than 1 to $5\,\mu$, and often much less, the above calculations suggest that nearly all of the surface area measured by nitrogen adsorption techniques is equally accessible to the dissolution process. We should also note that we have deliberately chosen values of k_2, S, and C_{sat} which would tend to maximize the possibility of dissolution *inhibition*. Where conditions of lower temperature, pH, salinity, and specific surface area occur there would be even less likelihood of interference. Values for k_2 for a wide variety of assemblages reported in Hurd and Theyer (1975) cluster around an average value of 10^{-7} cm^{-1} s^{-1} at 25°C implying that the above calculations are correct; however, the definitive experiment to test these calculations has yet to be done. The values of k_2 obtained from the dissolution of a three order of magnitude range of specific surface area samples should be compared and, further, the range of specific surface areas within a given sample must be narrow ($\pm 10\%$ or less) and the *total* area of sample in each dissolution experiment must initially be the same.

Although experiments to determine the variability of k_2 as a function of pH, temperature and ionic strength have been done separately or in pairs (Wirth and Gieskes, 1979; O'Connor and Greenberg, 1958; van Lier *et al.*, 1960; Hurd, 1973), a complete study done on all three variables using the same sample, especially for biogenic silica is not available. It does appear, however, that these variables act somewhat independently of one another in their effect on k_2. Thus a temporary, empirical approach using trends in the data from the above workers may be used until a proper data set is obtained.

So, analogous to the work of Wirth and Gieskes (1979), we shall begin with a reference point and use the ratio of rate constants under various conditions of temperature salinity and pH to estimate k_2 for various sets of conditions. If our reference point is pH 7.0, salinity $35^0/_{00}$ ($I \sim 0.7$), 25°C for which k_2 is ca. $4 \pm 1 \times 10^{-8}$ cm s^{-1}, and we wish to estimate k_2 at ca. 0°C, salinity $0.35^0/_{00}$ ($I \sim 0.007$), pH 5, then the following approximately obtains from the ratio estimates in Table 6.1:

$$k_2, 0°C, 0.35^0/_{00}, \text{pH } 5 = (k_2, 25°C, 35^0/_{00}, \text{pH } 7)\left(\frac{k_2, 0°C}{k_2, 25°C}\right)$$

$$\left(\frac{k_2, 0.35^0/_{00}}{k_2, 35^0/_{00}}\right)\left(\frac{k_2, \text{pH } 5}{k_2, \text{pH } 7}\right) \sim (4 \pm 1 \times 10^{-8} \text{ cm s}^{-1})\left(\frac{4 \pm 1 \times 10^{-9}}{4 \pm 1 \times 10^{-8}}\right)$$

$$\left(\frac{8 \pm 1 \times 10^{-9}}{4 \pm 1 \times 10^{-8}}\right)\left(\frac{2.5 \pm 3 \times 10^{-9}}{4 \pm 1 \times 10^{-8}}\right) \sim 5 \pm 3 \times 10^{-11} \text{ cm s}^{-1}$$

If we compare warm, salty, high pH surface waters of Equatorial waters with cold, fresh acidic lake waters, there may be as much as four orders of magnitude difference in the overall values of k_2, whereas solubility changes only by a factor of about 2.

Comprehensive models involving river–seawater mixing processes as well as those dealing with the settling of material through the thermocline *must* take into account the changes in k_2 and C_{sat} as approximated above.

For example, the slowing of dissolution rate in decomposing phytoplankton may not only be caused by surface blockage by organic matter and clumping, but also a lowering of pH in microenvironments by bacterial degradation processes. Only half a pH unit lower value would be necessary to decrease dissolution rates by a factor of 2.

The above method for estimating the overall trends in k_2 for a wide range of environments is a useful beginning but applies primarily to acid-cleaned material. Recently, Lawson, Hurd, and Pankratz (1978) reviewed a number of previous investigators' work on the dissolution rates of somewhat naturally decomposing biogenic silica in sea water at various temperatures. Figure 6.6 summarizes our k_2 estimates of all these experiments. Lawson et al. (1978) also produced a series of predictive equations for the dissolution rates of naturally decomposing biogenic silica as a function of temperature in sea water in the manner below (which will eventually also be used for a wider variety of environments when a suitable data set is available).

Dissolution rates are internally consistent with respect to changes in temperature. That is, the product of the temperature and the time required for a given weight per cent to dissolve is relatively constant for the range of temperatures studied. This suggests that for a salinity of $33 \pm 3^0/_{00}$ and in the pH range 7.9–8.3, predictive equations can be made for the rate constant k_2' and the solubility C_{sat} as functions of temperature using the Arrhenius equation and the Gibbs free energy equation (Daniels and Alberty, 1961). Thus

$$k_2 = fe^{-E_{act}/RT} \qquad (5)$$

where k_2 is the dissolution rate constant in cm s^{-1}, f is the frequency factor in cm s^{-1}, E_{act} is the energy of activation in cal mol^{-1}, R is a constant, 1.987 cal degK^{-1} mol^{-1}, and T is the temperature in degrees Kelvin. Using data in Table 6.1 between 28° and 11°C, we derive the following relationship:

$$k_2' = (41.6 \pm 2.2)e^{-13\,800/RT} \text{ cm s}^{-1} \qquad (6)$$

TABLE 6.1   Approximate ratio matrix dissolution rate constant k_2 for acid-cleaned biogenic silica.

T, °C	Ratio	Salinity %	Ratio	pH	Ratio
30	1.5			8.5	5.5
25	1.0	35	1.0	8.0	4.0
20	0.6	5	0.5	7.5	2.0
15	0.35	1	0.25	7.0	1.0
10	0.21			6.5	0.5
5	0.13			6.0	0.25
0	0.07			5.5	0.13
				5.0	0.0625

for our experiments where k_2' refers to the naturally decomposing phytoplankton studied. The energy of activation for this particular process is 13 800 cal mol^{-1}.

Further,

$$C_{sat} = e^{-\Delta G/RT} \tag{7}$$

where C_{sat} is the solubility of acid-cleaned biogenic silica at a given temperature and pH range (7.9–8.3), ΔG is the change in Gibbs free energy in cal mol^{-1} and the other terms are as before. Using data from Hurd and Theyer (1975)

$$C_{sat} = e^{-3830/RT} \times 10^{-3} \tag{8}$$

The 10^{-3} term converts from mol l^{-1} to mol cm^{-3} so that consistency of units is maintained.

Now, when $C_{sol} \ll C_{sat}$ the surface flux can be predicted as a function of temperature by multiplying equations (6) and (8) to give the initial or maximum flux:

$$\text{Flux}', \text{ mol cm}^{-2}\text{ s}^{-1} = (41.6 \pm 2.2 \times 10^{-3})e^{-17\,600/RT} \tag{9}$$

This when multiplied by the molar volume (~ 26 cm^3 mol^{-1}, assuming a silica density of 2.3 gm cm^{-3}) gives the rate at which the nominal radius of a solid particle of silica initially decreases as a function of temperature for decomposing phytoplankton from Kaneohe Bay:

$$\text{linear dissolution rate (LDR}'), \text{ cm s}^{-1} = (1.08 \pm 0.05)e^{-17\,600/RT} \tag{10}$$

It is important to note that although the units of equations (6) and (10) are the same, the physical meaning of the two equations is quite different. Whereas the units of equation (6) arise simply to balance the units in the other terms of equation (1), those units in equation (10) actually represent the rate at which the radius of a given non-porous biogenic silica particle decreases under a given set of conditions. The numerical difference between these two equations for a given temperature is approximately four orders of magnitude. That Lerman and Lal (1977) appear to have chosen values of equation (6) rather than equation (10) suggests the possibility of a four-order-of-magnitude error in their calculations.

It is entirely possible that the dissolution rate equations describing our experiments (equations 6, 9, and 10) may underestimate the rate of biogenic silica dissolution above 10°C compared with previous investigators' work (see Fig. 6.6). For this reason we offer a different value of k_2' (called k_2'') which averages those workers' data:

$$k_2'' = (2.6 \times 10^{12})e^{-27\,300/RT} \tag{11}$$

and gives a different value of E_{act} of $-27\,300$ cal mol^{-1}. Equation (11) may then be treated analogously to equation (6) to a given surface flux

$$\text{Flux}'',\ \text{mol cm}^{-2}\,\text{s}^{-1} = (2.6 \times 10^9)e^{-31\,130/RT} \tag{12}$$

and a linear dissolution rate, LDR''

$$\text{LDR}'' = (6.9 \times 10^{10})e^{-31\,130/RT} \tag{13}$$

Figure 6.7 graphically demonstrates the estimated effects of LDR' (solid line) and LDR'' (dashed line) for a variety of particle sizes. The stippled area for each particle half-thickness represents the realistic range of possible times required for that particle to dissolve under natural conditions in the water column. Dissolution rates are much slower below the sediment–water interface as argued by Hurd (1973).

Within the last few years, a particularly interesting method for estimating *in situ* dissolution rates has appeared (Nelson, 1975; Nelson and Georing, 1977a,b). In this approach natural populations of phytoplankton are

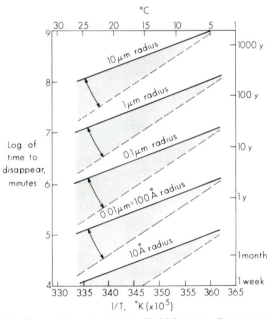

FIG. 6.7   Log of time for particles of various half-thickness to disappear as a function of the reciprocal of absolute temperature. The solid lines are from data in Lawson, Hurd, and Pankratz, 1978, and the dashed lines are a summary of previous worker's results. The stippled areas are meant to suggest that all values between the two sets of estimates are equally valid and depend on the conditions under study (figure from Lawson, Hurd, and Pankratz, 1978).

sampled from various layers within the euphotic zone, the sample solutions enriched with ^{30}Si and each incubated at a light intensity corresponding to its depth of sampling. Rationale, technique, and additional basic equations are well described in the above references. The dilution of the ^{30}Si-enriched solution as a result of the dissolution of siliceous frustules which are rich in ^{28}Si is followed and what are termed absolute and specific dissolution rates are produced as a function of depth (i.e. light intensity). The absolute dissolution rate, ρ_{dis}, is defined as:

$$\rho_{dis} = (Si(OH)_4)_t \frac{(^{28}Si_f - {}^{28}Si_i)}{(^{28}Si_n - {}^{28}Si_i)} (\Delta t)^{-1} \qquad (14)$$

where $(Si(OH)_4)_t$ is the total silicic acid concentration (ambient plus label) of the seawater sample before incubation, $^{28}Si_i$ the atom % ^{28}Si initially, $^{28}Si_f$ after incubation for time Δt, and $^{28}Si_n$ the natural atom % in the silica skeletons, which is assumed to be 92.18. The units of ρ_{dis} are mol vol^{-1} time^{-1}. The specific dissolution rate, V_{dis}, having the units of time^{-1} is defined as:

$$V_{dis} = \rho_{dis}/P-Si \qquad (15)$$

where ρ_{dis} is defined as before and P–Si is the concentration of particulate silica in mol vol^{-1}. V_{dis} is the moles of dissolved silica produced per mole of particulate silica per unit time and is similar to a percent dissolution rate per unit time.

Nelson (1975) has suggested that equation (1) is related to (15) in the following manner:

$$V_{dis} = k_2(C_{sat} - C_{sol})S/P-Si \qquad (16)$$

and that if pH, temperature and salinity were held constant (as in their incubation experiments) that the major variable would be the effective area per mole particulate silica per unit volume of solution.

In Fig. 6.8 Nelson's (1975) Fig. 1 of V_{dis} versus surface area per gram of silica (specific surface area) is replotted. Note that moles of silica per volume have been converted to grams per volume by dividing by 60. It appears that Nelson did this correctly in his calculations but omitted the number in the figure and text explanation. Plotted on Fig. 6.8 are Nelson's estimates of V_{dis} from previous worker's experiments (for details see Fig. 6.8 caption). The circles show naturally decomposing material at various temperatures, the squares are acid-cleaned values and the triangles are Nelson's (1975) dissolution estimates from several live cultures in exponential growth stages.

If the dissolution rate constant varies simply with temperature and the available area is initially the same for each experiment, then the data distribution should be essentially vertical, which it generally is for Lewin's acid-cleaned material (squares, N. peliculosa). The solutions were unstirred and as such the estimated surface areas are slightly lower than those actually measured by nitrogen adsorption. However, none of those diatoms "killed but not acid-cleaned" generally follow the vertical trend, although all dissolve more slowly than the "acid-cleaned" subsamples of the same suspension. It may be that bacterial degradation occurs with increasing speed and efficiency at higher temperatures thereby lowering the pH near the frustule and reducing the value of k_2. Nelson's values range from the "killed but not acid-cleaned" values to acid-cleaned values, depending on what is believed to be the actual specific surface area of the skeletal material, as well as the effective area available for dissolution. Data from a field experiment (Nelson and Goering, 1977a) are also plotted in the temperature range estimated to be 15° to 20°C, as the various hatchings and cross-hatchings show (see Fig. 6.8 caption for more details). There is a two order of

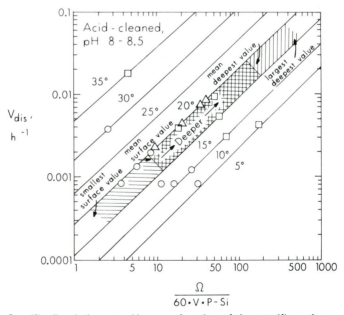

FIG. 6.8   Specific dissolution rate, V_{dis}, as a function of the specific surface area of the biogenic silica (after Nelson, 1975, Fig. 1) for a range of temperatures. Open circles = naturally decomposing phytoplankton; open squares = acid-cleaned material; and open triangles = estimates by Nelson (1975) done on living cultures. The hatched and cross-hatched areas summarize data from Nelson (1975) also shown in Fig. 6.9, giving estimates of V_{dis}, for a field study off West Africa. N.B. solid lines were generated using acid-cleaned values for each temperature, suggesting that Nelson's deepest samples are dissolving very quickly and may have very high specific surface areas.

magnitude range in the V_{dis} values and a slightly more than order of magnitude range when comparing their surface (i.e. 100% light level) to deepest (0.1% light level) values. Figure 6.9 shows the same field data now plotted versus per cent light level and against the ratio of surface light to light below the surface at various depths. The data were plotted this way because although the samples were collected at various depths, all the experiments were run at surface water temperatures onboard ship, and depth was simulated by decreasing light levels with neutral density filters. There is an obvious overall direct correlation between light attenuation and dissolution rate as measured by this method. This may mean that as fast as the original skeletons dissolve, the diatoms remove much of that material from solution at high light levels, whereas at low light levels there may not be sufficient energy available to keep up with dissolution. In the end the actual exchange

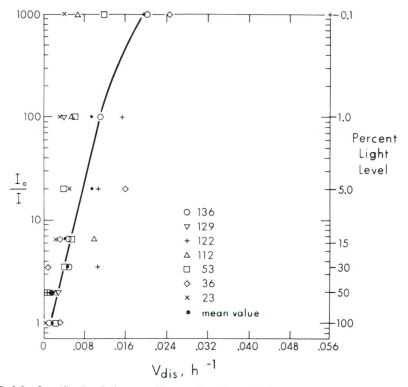

FIG. 6.9   Specific dissolution rate, V_{dis}, as a function of both percentage light level and the ratio of surface light intensity, I_o, to light at various attenuations, I. N.B. Dissolution rate appears to be strongly correlated with light intensity, although there is a great deal of scatter (data from Nelson, 1975).

rates in mass/area time are still unknown because the areas are not really known. There are implications that quite high areas may be involved. From the position of the largest V_{dis} value (0.056 h^{-1}), if the silica were dissolving at the acid-cleaned rate then their specific surface areas could be as high as 400 to $500 \text{ m}^2 \text{ g}^{-1}$.

It is obvious at this point that in addition to the above useful field studies, a number of basic laboratory studies need to be done in order to relate this particular method to work done previously:

1. To see if the exchange rates are in fact comparable to acid-cleaned silica, several suspensions of known specific surface area acid-cleaned material and thus *known* S/P–Si should be run at well-defined temperature and pH conditions to see if V_{dis} and k_2 are in fact simply related.
2. A culture in exponential growth stage should be tested over a much wider range of time scales and/or weight per cent exchanges. It is not possible to know whether V_{dis} is a function of per cent exchange using only a single time interval as has been done thus far.
3. Splits of cultures treated in various ways should also be monitored, i.e. (a) one split treated alive; (b) one split in darkness; (c) one split frozen and resuspended; (d) one split killed by glutaral-dehyde addition; (e) one split acid-cleaned and resuspended, etc.

Most of these approaches have been used singly by investigators in the past. Nelson now has the excellent opportunity to tie all this data together and relate it to his existing field programme in a meaningful manner.

6.2.3 Questions Relative to the Dissolution Rate Section

1. How does the dissolution rate of the frustule of a diatom or radiolarian vary during its lifetime and after its death (on a mass/area time basis)?
2. How do particle size and pore size actually affect dissolution rate?
3. Why do porous layers result during the dissolution process? Is the infilling material substantially different than the remaining material? Is it mathematically possible and useful to describe the dissolution process in terms the relatively rapid formation of a porous layer which then itself later dissolves? How rapidly does the porous layer formation occur relative to the various oceanographic and limnological time scales and relative to silica recycling within a particular system?
4. Are the effects of pH, temperature, and salinity in fact separate from one another or are there synergistic effects?
5. What is the effect of internal hydroxyl content on solid dissolution rates? In that the internal hydroxyl content of vitreous silica must be less than that of low-temperature precipitated biogenic silica could this be one of the reasons why vitreous silica dissolution rates are slower?

6.3 DENSITY AND REFRACTIVE INDEX DIAGRAMS FOR SILICA-WATER POLYMORPHS

Frondel (1962) and Sosman (1965) thoroughly reviewed and organized the work regarding the density and refractive index of various silica polymorphs. The data in Table 6.2 and many of the ideas presented in this paper have drawn heavily on these two sources. The range of properties for inorganically precipitated opals (Taliaferro, 1935; Kokta, 1931, as cited in Taliaferro), biogenically precipitated silica (this work), water, and ice have also been given. Figure 6.10 shows the essentially linear relationship between refractive index and density for these polymorphs. Since density can normally be measured to 0.01 g cm^{-3}, and refractive index to 0.005, gross changes in the transition sequence from amorphous silica to quartz, or amorphous silica to cristobalite, should be relatively easy to observe. In addition, these properties are attractive for investigators since quite small amounts of sample may normally yield reasonably high precision and accuracy without difficulty.

Earlier workers (Taliaferro, 1935; Kokta, 1931) have tried to explain the properties of inorganically precipitated opals or hydrogels of silica and water by comparing the weight per cent water, and either specific gravity *or* refractive index with theoretical curves. These curves were generated by mixing various *volume* percentages of water and a property of either vitreous

TABLE 6.2 Selected properties of silica and water polymorphs and mixtures.

	Density g cm^3	Molar volume cm^3 mol^{-1}	Refractive Indices			
			ε	ω		Biref.
Quartz	2.65	22.60	1.553	1.544		0.009
			ε	ω		Biref.
low Cristobalite	2.33 ± 0.01	25.8 ± 0.1	1.487	1.484		0.003
			α	β	γ	Biref.
low Tridymite	2.26 ± 0.01	26.6 ± 0.1	1.469-73	1.469	1.473-4	0.003
Silica Glass (Lechateliarite)	2.19 ± 0.02	27.4 ± 0.2	1.458			
Melanophlogite	1.99 ± 0.01	30.2 ± 0.1	1.425 ± 0.002			
Inorganically precipitated opal	1.98–2.16	—	1.441–1.459			
Radiolarians	1.7–2.05	—	1.400–1.448			
Diatoms	1.98–2.01	—	1.4409–1.4413			
Sponge Spicules	1.99–2.04	—	1.442–1.467			
Water	1.00	18.0	1.333			
Ice	0.917	19.6	1.303			

silica or low cristobalite. While such techniques are not inherently incorrect, they tend to obscure inconsistencies between specific gravity or density and refractive index for a given water content; it is possible that much more information may be gained by plotting the properties of a given sample on a refractive index versus density diagram, having isolines of constant weight-percent water, since the inconsistencies of samples thus become readily apparent.

The following formulae have been used to estimate those properties produced by the mixing of various silica polymorphs with each other and with water of varying densities.

$$\text{Density of mixture} = \frac{(M_B - M_A)x_B + M_a}{(V_B - V_A)x_B + V_A} \tag{17}$$

where M_A and M_B are the molecular weights of substances A and B in g mol^{-1}; V_A and V_B are their respective molar volumes in cm^3 mol^{-1}; and x_B is the mole fraction of substance B.

The refractive index of the mixture is obtained by first using the molar refraction formula to obtain the molar refraction R, in cm^3 mol^{-1} of each substance:

$$R = \frac{(n^2 - 1)}{(n^2 + 2)} \cdot \frac{M}{d} \tag{18}$$

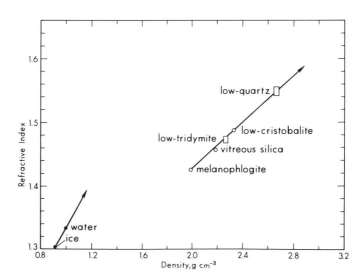

FIG. 6.10 General relationship among water and silica polymorphs; most of the subsequent figures are enlargements of this figure, from Hurd and Theyer (1977).

where n is the refractive index; M the molecular weight of the substance in g mol^{-1}; and d is the density in g cm^{-3}. Then, since the fraction M/d equals the molar volume V, one can alter equation (17) slightly to give:

$$\frac{(n_{mix}^2 - 1)}{(n_{mix}^2 + 2)} = \frac{(R_B - R_A)x_B + R_A}{(V_B - V_A)x_B + V_A} \tag{19}$$

where n_{mix} is the refractive index of the mixture, and the other terms are as before. It is important to note that for equations (17), (18), and (19) the determination of the true density of the substance fixes the value of the molar volume V and its variable depending on which polymorph in a given series is being considered. The values of M and R (within a few per cent) are fixed for a given series of polymorphs. Although R is dependent on the value of the refractive index as well as density, it can be averaged by using data from several polymorphs within a series and as such is not as subject to errors as V. The solution to the left-hand side of the equation can be solved by referring to a set of tables such as those given in Batsanov (1961) or by computor iteration and substitution techniques. Figure 6.11 gives an example of such a diagram, where water of density 1.00 has been mixed with low cristobalite and melanophlogite. The mixing curve for air and low cristobalite is also

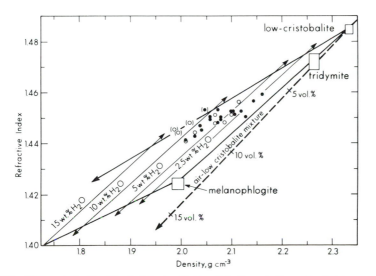

FIG. 6.11 Computed refractive index-density combinations to be expected upon mixing water (n = 1.333, d = 1.00 g cm^{-3}) in various weight percentages with different silica polymorphs. The open circles are data of Taliaferro (1935), the closed ones of Kokta (1931). Additional dashed line shows the properties of an air-low cristobalite mixture, from Hurd and Theyer (1977).

given, along with isolines for weight per cent water. Since the structure, and therefore density, of ice has been suggested for a number of hydrated materials (Moerman, 1938, as cited in Hendricks and Jefferson, 1938; Patrick, 1951 as cited in Iler, 1955), a diagram for ice similar to Fig. 6.11 may also be prepared.

When the data of Taliaferro and Kokta are plotted on such diagrams, both water and ice yield essentially the same agreement (correlation coefficient = 0.97) between estimated and determined water content. That is, the refractive index and density of a given sample are plotted on Fig. 6.11, and the weight per cent water estimated by the position of the point relative to the weight per cent water isolines; then this number is compared with the laboratory-determined ignition losses which are assumed to be only water.

However, when an attempt is made to estimate the properties of the silica phase by subtracting out the water phase, water of ice density produces tridymite-like properties (density = 2.267 ± 0.025, n = 1.473 ± 0.003), whereas water of density 1.00 produces silica properties nearer to those of vitreous silica (density = 2.237 ± 0.021, n = 1.468 ± 0.003). This suggests that the actual density of water in the opal structure is important to determine, if we are going to use these types of subtraction methods for estimating the properties of the silica phase.

The regression line for refractive index versus density (excluding

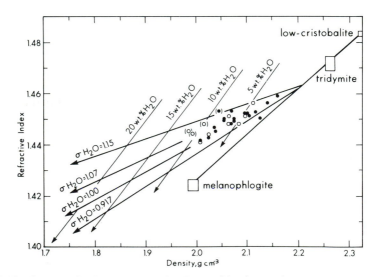

FIG. 6.12   Computed refractive index-density combinations to be expected upon mixing water of varying densities and refractive indices with a silica polymorph having the properties of silica glass (n = 1.465, d = 2.21 g cm⁻³).

Taliaferro's samples I, II, and X and Kokta's sample 8 because of obvious textural irregularities) is:

Refractive Index = (0.1071 ± 0.0075)(Density) +
$$1.229(\text{correlation coefficient} = 0.92) \quad (20)$$

This line intersects our suggested silica polymorph line at density 2.21 g cm^{-3}, refractive index 1.464. Furthermore, this line is distinctly non-parallel to a line from the same point using water of either 1.00 or 0.917 g cm^{-3} density. If the density of the water in the opal is increased to 1.07 g cm^{-3}, however, much better agreement results. Figure 6.12 shows a diagram similar to Fig. 6.11, except that several different lines using water of varying density have been generated from the above regression line intersection point.

Unfortunately, a number of inconsistencies and possibilities for contamination exist which further complicate the direct application of simple mixing models to the data of Taliaferro and of Kokta.

1. Whereas the regression line we have suggested gives average values of 2.21 g cm^{-3} for density and 1.465 for refractive index at zero water content, regression lines of weight per cent water versus density or refractive index separately yield different values:

Wt. per cent water = $77.55 - (34.34 \pm 3.49)$(density) $\qquad$ (21)

Wt. per cent water = $630.1 - (430.4 \pm 59)$(refractive index) $\qquad$ (22)

(correlation coefficients for above, and -0.74, -0.87, respectively)

According to equations (21) and (22) at zero water content, the sample would have density of 2.26 ± 0.23 and a refractive index of 1.464 ± 0.199. The large error associated with equations (21) and (22) suggests that our estimates of 2.21 and 1.465 are possible, but that considerable variation near these figures exists.

2. Another possibility which might cause deviation from the simple silica–water mixing curves is the presence of gas vacuoles within the samples. Taliaferro suggested that some samples contained minute vacuoles filled with air or as a vacuum (his samples I, II, and X and perhaps Kokta's sample 8). If we can use a simple volume per cent mixing formula to predict the properties of a silica–air mixture, then, assuming a density of zero for air and a refractive index of 1.000, only 2 volume per cent of air in these naturally occurring samples is necessary to give the approximate density and refractive index values of low tridymite to a low cristobalite sample.

The author does not know what chemicals are used in the manufacture of various refractive index oils; however, it appears entirely possible that microstructures will exist throughout the opal structure that will allow the entrance of O_2, N_2, or H_2O but exclude the presence of somewhat larger molecules which must make up these oils. This would give rise to a systematic underestimate of the refractive index with an unknown degree of random error. Assuming that the densities of the opals were determined in a pycnometer, using water as the displacing fluid, and that the sample was either boiled or thoroughly evacuated to exclude air from its pores, then closer approximations of the true densities would result. When determining refractive index, however, samples are rarely, if ever, evacuated to remove air because the solids studied are assumed to be nonporous.

6.3.1   Application of the Density-Refractive Index Plot to Biogenic Silica

Figures 6.13, 6.14, and 6.15 show the data from Appendix A plotted on the suggested density-refractive index diagrams. These samples have been somewhat arbitrarily divided into three groups with the solubility ranges 1500–1750 µM, 1000–1485 µM, and 100–800 µM at 25°C (Hurd and Theyer, 1975). The divisions could have also been specific surface area or even age ranges. Also included in each figure are the combined data of Taliaferro (1935) and of Kokta (1931) for reference.

The mean density value of the various species in an assemblage when compared with the bulk density value for the same assemblage (samples 7, 18, 21, 22, 23, 24, 25, 26, 28, and 34) showed generally good agreement

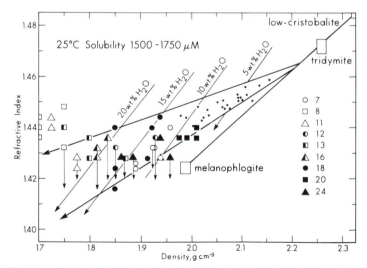

FIG. 6.13   Observed refractive index-density combinations for those radiolarian as-semblages having the 25°C solubility range 1500–1750 µM Si(OH)$_4$. Number-symbol pairs refer to I.D. numbers in Hurd and Theyer (1977) (figure from Hurd and Theyer, 1977).

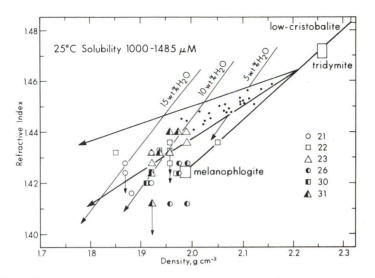

FIG. 6.14 Observed refractive index-density combinations for those radiolarian assemblages having the 25°C solubility range 1000–1450 µM Si(OH)$_4$. Number-symbol pairs refer to I.D. numbers in Hurd and Theyer (1977) (figure from Hurd and Theyer, 1977).

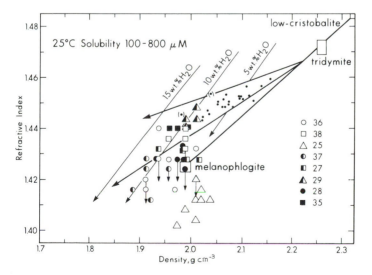

FIG. 6.15 Observed refractive index-density combinations for those radiolarian assemblages having the 25°C solubility range 100 to 850 µM Si(OH)$_4$. Number-symbol pairs refer to I.D. numbers in Hurd and Theyer (1977) (figure from Hurd and Theyer, 1977).

(± 0.03 g cm^{-3}) except for samples 18 and 28 which had much higher bulk densities, possibly because of heavy mineral contamination. In some cases where a wide range of refractive indices was found for a given species, the extent of the range is indicated by an arrow extending downward from the highest value.

Only about one-third of the data appears to follow the trends shown by the inorganically precipitated opals studied by Taliaferro (1935) and Kokta (1931) (that is, samples 7, 12, 18, 21, 22, 23, 24, 29, 31, 36, and 37). That is, for a given combination of density and refractive index, the water content of the bulk sample, estimated by ignition loss agreed within about 3% absolute with that estimated by its position on the diagram.

Those samples greatly to the left of their ignition loss estimated values (samples 8, 11, 13, and 16) generally have both high solubility and high specific surface area; those samples greatly to the right of their estimated water content (samples 20, 25, 26, 27, 28, 30, 36, and 38) have generally much lower specific surface areas and solubilities, suggestive of both morphological and diagenetic change. The authors again wish to stress the importance of determining the nature of the porous structure of these solids, since it is felt to be the most important factor causing the patterns in Figs 6.13, 6.14, and 6.15.

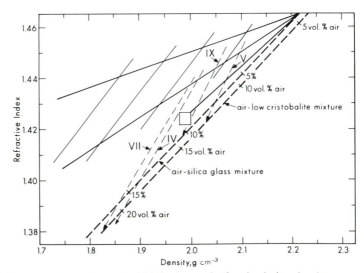

FIG. 6.16 Several of Taliaferro's (1935) observed refractive index-density measurements made before and after dehydration. N.B. approximate parallelism between the dashed lines showing dehydration paths and the solid lines for constant weight per cent water; the difficulty of accurately estimating the volume percentate of air present after heating if the density of the remaining solid phase is not known accurately (figure from Hurd and Theyer, 1977).

Figure 6.16 shows calculated paths resulting from gradual water loss *without* structural volume change, as postulated by Taliaferro (1935). Also shown are several endpoints for his actual dehydration experiments. Readers should note that the slopes of these lines essentially parallel nearby weight per cent water lines. It is possible that when some of the samples were first dried at 105° to 110°C, water was lost from certain small-necked pores which were not readily re-filled by the index oils or high density liquids. This appears to be the case for portions of samples 21, 28, 30, 31, 36, and 37. Such a model still does not satisfactorily explain why some samples give normal indices and low densities (8, 11, 13, 16).

Figures 6.17 and 6.18 show weight per cent water versus density and refractive index, respectively. In both diagrams the solid line represents the regression line using equations (21) and (22) and in each case is based only on the combined data of Taliaferro and of Kokta, whose data are shown by the small closed circles. Diagonal shading shows samples whose properties were to the left of their estimated water content position in Figs 6.13, 6.14, and 6.15; cross-hatched shading shows samples to the right, and unshaded samples are those with good agreement. In Fig. 6.17, the larger circles represent bulk sample densities done using a pycnometer; half-shaded circles are those for which only water content and density information were available. The rectangles represent the best estimate of the range of densities as estimated by simply averaging all the values of a given assemblage and including a reasonable range of values based on the estimate of the standard

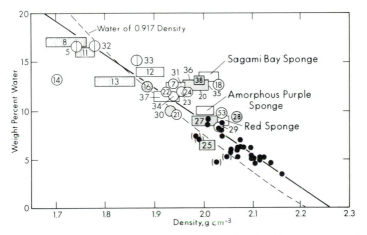

FIG. 6.17. Observed weight per cent water-density data for biogenic silica (Hurd and Theyer, 1977) and inorganically precipitated silica (data of Taliaferro, 1935 and Kokta, 1931) (symbols explained in text, figure from Hurd and Theyer, 1977).

deviation of this mean. Likewise in Fig. 6.18, the upper and lower ranges of values were estimated by averaging the higher and lower values separately and using those averages to bracket a given estimate. It is clear that such estimates are a crude but justifiable beginning in the absence of the actual water content data for individual species.

In Fig. 6.17, the general agreement between the trend of the inorganic opals and that of the biogenic opals is quite good. *If* one can regard the inorganic opal analyses as being free from inconsistencies, then the density measurements of biogenic opals appear nearly as trouble-free. Examination of Fig. 6.18, however, shows a great deal of scatter in the biogenic silica samples as compared with the inorganic samples. Furthermore, whereas it would appear reasonable to suggest that values below the regression line result from certain cracks or pores not being filled by the index oils, one must now wonder if it is possible that certain biogenic samples can actually condense the oils to such an extent as to increase the measured index of the sample. The lower index ranges of portions of samples 8, 11, 13, and 16 suggest this possibility, but we have been unable to find corroborative evidence of this phenomenon in the literature.

Because of the variability of surface coverage by hydroxyl groups and presumably water molecules as well, the unknown degree of pore size and variability, and pore volume as a function of geologic age, water content, and species type, and the possible interaction of index oils with the solid surface, it appears that any estimate of the refractive index or density of the silica phase

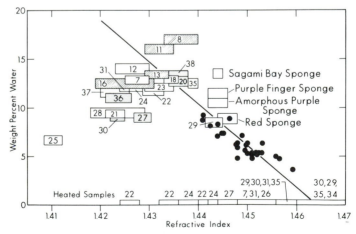

FIG. 6.18 Observed weight per cent water–refractive index data for biogenic silica (Hurd and Theyer, 1977) and inorganically precipitated silica (data of Taliaferro, 1935, and Kokta, 1931) (symbols explained in text, figure from Hurd and Theyer, 1977).

in biogenic silica can have a large (about 10%) error associated with it. This does not mean that such measurements cannot yield useful information, but rather that additional data such as the NMR studies of Segnit *et al.* (1965), or density determinations using gas displacement rather than liquid may be necessary before significant improvement in the accuracy of these numbers can occur.

6.4 SPECIFIC SURFACE AREA AND PORE VOLUME OF BIOGENIC SILICA

During the past 15 years, transmission electron microscopy (TEM) research dealing with the precipitation of silica by diatoms and sponges (Drum and Pankratz, 1964, Fig. 4; Stoermer *et al.*, 1965, Fig. 8; Dawson, 1973; Simpson and Vaccaro, 1974) has shown that the surface of newly formed frustule material can be quite porous.

Considering the surface of a newly forming diatom frustule as being a collection of loosely cemented, small particles, then the size of these particles may be approximated using specific surface area measurements. A range of 95 to 139 $m^2 g^{-1}$ suggests mean spheroid radii of about 140 to 95 Å. But when the specific surface area of a substance is measured and expressed as area per unit weight of that substance, it is important to remember that the solid studied may be a collection of particles having a wide size spectrum. One may calculate an *average* particle size, using the specific surface area data and the density of the solid (Iler, 1955, pp. 103 and 242), but this will be a *mean* or effective particle size which may or may not actually exist in the solid. Thus the mean particle size estimates should be useful as a better-than-other-of-magnitude estimate when combined with the LDR' or LDR" to estimate how long that substance will take to dissolve completely under a given set of conditions, given the proviso that that portion of the sample which is made up of quite small particles may well disappear more rapidly than the larger particle portion.

What follows is a partial discussion of recent surface area and pore volume work done on a series of Recent through Eocene age siliceous skeletons. For more detail and a description of methods and materials see Hurd *et al.* (in press).

Earlier work (Hurd and Theyer, 1975) suggested trends of decreasing specific surface area with increasing age of siliceous skeletons but was based on a relatively few number of samples. Figure 6.19, based on previous samples as well as another three dozen samples from Deep-Sea Drilling Project sites 65, 70, and 71 shows that the trend suggested earlier still holds, i.e. in general, older samples have much lower specific surface areas than younger ones, and that there is a relatively large change in specific surface area during the last 5 to 7 mybp (millions of years before present).

More can be learnt about the particles and their spacing by first determining the volume of interparticle space per gram of material (specific pore volume, $cm^3 g^{-1}$) and relating that space to the surface area of the particles to obtain an estimate of the average half-distance between particles. Such information is useful in that it may give clues regarding the precipitation of silica by the organisms, the settling rate of the skeleton through the water column and the preservation of the skeleton in the water column and sediments.

In Fig. 6.20 the specific pore volume in $cm^3 g^{-1}$ as determined by nitrogen desorption at 0.95 p/p_0 N_2 in He (see Hurd *et al.*, in press, for details) on the same samples as in Fig. 6.19 are plotted as a function of age. The same trends appear for specific pore volume as for specific surface area: relatively large pore volumes occur during the last 5–7 mybp decreasing to almost negligible values in the late Eocene (40 mybp). Specific pore volume data may be converted to porosity data by assuming that the density of the solid phase is approximately 2.2 $g\ cm^{-3}$ and using the equation:

$$\% \text{ Porosity} = \frac{V_{pores}}{V_{pores} + V_{solid}} \times 100 \tag{23}$$

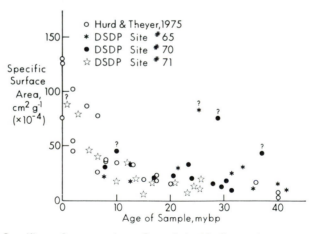

FIG. 6.19 Specific surface area in $cm^2\ gm^{-1}$ ($\times 10^{-4}$) vs. the age of the siliceous assemblage from the sediments in millions of years before present (mybp). The general properties of the assemblages alter rapidly during the time period Recent to 10–15 mybp (mid-Miocene) and remain relatively constant thereafter. Question marked values contained obvious amounts of contaminants but were analysed to show the effects of these contaminants on sample values (figure from Hurd *et al.*, in press).

where V_{pores} is the specific pore volume in $cm^3 \ g^{-1}$ and V_{solid} is the reciprocal of the density of the solid, $0.45 \ cm^3 \ g^{-1}$. Both V_{pores} and V_{solid} values are approximate. The limitations of the specific pore volume determination exclude pores with apertures greater than 250–500 Å in radius on the outer surface of the solid because of the gas mixture we used, thus it is possible that we may not be measuring all the pores within the porous structure; from this point of view V_{pores} is a lower limit estimate. It should be noted, however, that if pores larger than 250–500 Å radius exist some distance within the solid that they would be included in our measurements because we first saturate the solid and then equilibrate at either 0.95 or 0.90 p/p_0. If a layer of smaller pore lies outside the larger pores, then the larger pores will probably not empty at the lower p/p_0 value.

On the basis of transmission electron microscope data, to be discussed later, it also appears that portions of the skeleton may have closed pores which may contain water or another substance which have the effect of lowering the solid density; the above estimate of V_{solid} may be slightly high for this reason. The data in Fig. 6.20 converted from specific pores volume to per cent porosity are shown in Fig. 6.21 for reader convenience, but the above limitations on the accuracy of V_{pores} and V_{solid} suggest that the porosity estimates have an error of about $\pm 20\%$.

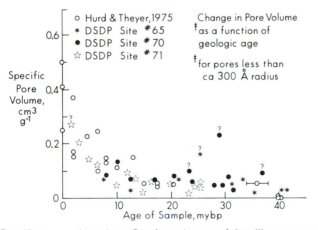

FIG. 6.20 Specific pore volume in $cm^3 \ g^{-1}$ vs. the age of the siliceous assemblage from the sediments in millions of years before present (mybp). The general properties of the assemblages alter rapidly during the time period Recent to 10–15 mybp and remain relatively constant thereafter. Those values having question marks by them contained obvious amounts of contaminants but were analysed to show the effects of these contaminants on sample values. Note that we are nominally measuring pore sizes up to ca. 250 to 500 Å in radius because of the gas mixture used in these determinations (95% N_2/5%He).

Figures 6.19–6.21 clearly show overall changes in the micro- or ultrastructural geometry of siliceous assemblages from various sediment types
with time. A decrease in specific surface area means that the average particle
size is increasing; a decrease in specific pore volume means that less total,
open-pore volume is available although there may still be considerable
closed-pore volume. It would be tempting to ascribe these effects solely to
diagenesis the infilling of pores by a more stable silica polymorph or some
type of aluminosilicate would result in both of the above effects and in earlier
work (Hurd and Theyer, 1975) decreasing solubility with increasing age of
sample has been observed. Micropaleontologists have also noted considerable changes in the overall skeletal morphology of radiolaria (Kling,
1978, and references therein) since the Eocene and have commented on
overall changes in skeletal structure (Harper and Knoll, 1975; Moore, 1969).
However, one must first see whether any consistent relationship exists
between specific surface area and specific pore volume before we can sort out
various effects.

If specific pore volume versus specific surface area is plotted, the slope of
the curve has the units of length. This measurement can be interpreted as a
mean pore radius or mean interparticle half-spacing and obtains from the
following equation:

$$r = \frac{2V}{\gamma A} \tag{24}$$

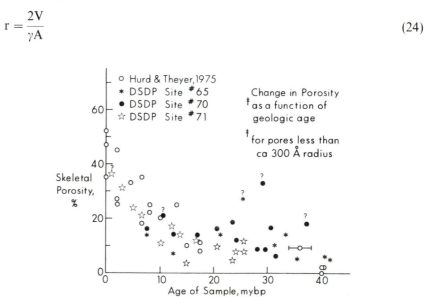

FIG. 6.21   Skeletal porosity in per cent vs. the age of the siliceous assemblage from the
sediments in millions of years before present (mybp). The same trends seen here exist for
Fig. 6.20 whose data were used to compute the porosities (Figs 6.20 and 6.21 from Hurd *et
al.*, in press).

where r is the mean pore radius in cm; V, the specific pore volume in $cm^3 g^{-1}$; A, the specific surface area in $cm^2 g^{-1}$; and γ (gamma) a shape factor depending on the geometrical shape and orientation of the particles and pores to one another. The derivation of the equation and a number of the possibilities and limitations for various particle and pore shapes have been extensively discussed by Everett (1958), Gregg and Sing (1965), Aristov *et al.* (1962), and Wade (1964, 1965).

In Fig. 6.22 several trends for a V vs. A diagram are schematically illustrated. A strictly vertical trend in the data, representing constant area and changing volume, could be achieved by beginning with a loosely packed powder and slowly compressing it (Carman and Raal, 1951; Zwietering, 1956; Kiselev, 1958; Wade, 1965). On the other hand, a horizontal trend in the data might suggest that somehow the distance between the particles remained constant while particle sizes were diminished or enlarged. In general, we might more usually expect to find neither extreme but somewhere in between. A diagonal trend might be generated (1) by the growth of particles and infilling of pores and/or (2) varying the ratio of a uniformly porous portion of the material to a uniformly non-porous portion. It is probable that both of the above occur in radiolarian and diatom skeletons and that (1) results predominantly from diagenesis while (2) may occur as a result of diatom and radiolarian physiological effects on skeletal formation and the subsequent dissolution of the skeleton. Preliminary data regarding the second statement will be presented at this time.

A portion of the data in Figs 6.19 and 6.20 have been plotted vs. one

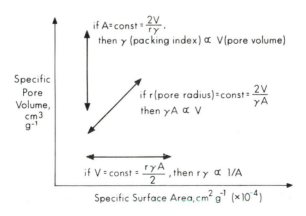

FIG. 6.22 Illustrative plot of specific pore volume in $cm^3 gm^{-1}$ vs. specific surface area in $cm^2 g^{-1}$ ($\times 10^{-4}$). The various trends shown are discussed in the text (figure from Hurd *et al.*, in press).

another in Fig. 6.23. Consideration of this figure prompts the following questions:

First, what is the overall significance of the value and variability of the slope (that is, the estimate of r) and of the range of A and V values shown in Fig. 6.23; and, assuming that we are interested in the shape and spacing of the particles from one another, as well as whatever changes may occur with time, what is a reasonable range of values for gamma?

The first point to notice is that the numerical value of r as derived from the slope is small, on the order of 35 Å (assuming an approximate value of 2 for gamma (see Hurd *et al.*, in press). A second point is that over nearly 2 orders of magnitude range in specific surface area and pore volume, the ratio of V/A does not generally vary by more than a factor of two and often, within a given core, by no more than 15 % around the mean value (see Hurd *et al.*, in press). This suggests that both the mean distance and/or the size distribution of

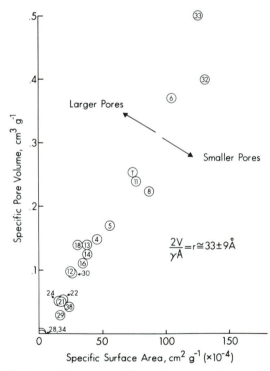

FIG. 6.23 Specific pore volume vs. specific surface area using a portion of the data from Figs 6.19 and 6.20, showing a linear relationship over a wide range of sample types and ages. Circled numbers refer to sample I.D. numbers in Hurd *et al.* (in press) (figure from Hurd *et al.*, 1979). Copyright 1979 by the American Association for the Advancement of Science.

distances between particles of biogenically precipitated silica seem to have remained relatively constant during the last 40 million years according to gas adsorption data.

The numerical value of r as estimated using equation (23) depends on what value for gamma is chosen. Everett (1958) has suggested that it may be difficult to choose an appropriate value for gamma without having some knowledge of the size and shape of the particles and pores because a number of different systems (sheets, rods, cubes, etc.) of varying geometry may still produce generally similar V/A plots. In addition, even if one knew the shape of the particles, there would surely be a distribution of both particle and pore sizes which would tend to smooth what might have been a better defined trend.

In order to determine more accurately which of Everett's models (if any) best described the physical properties of our system, high resolution (256 000 ×) transmission electron micrographs of thin sections of biogenically precipitated silica were taken (Hurd et al., 1979). Plates 6.1, 6.2, and 6.3 show several such high resolution shots of polycistine radiolaria of various ages. These plates clearly show that a great many small holes exist but that an accurate size distribution by visual means might be quite difficult.

The range of values for V and A presents another interesting problem. It was earlier suggested (Hurd et al., 1979) based on a model by Everett (1958) that the range of V/A values may be caused by mixing varying proportions of two end-members: one, a solid which is either non-porous or has closed pores which cannot be entered by nitrogen or water molecules, and another which is quite porous and has either a narrow range of pores and particle sizes or a consistent distribution of pores and particle sizes. Transmission electron microscopy has revealed the presence of such end-members and Plates 6.4, 6.5, and 6.6 show a number of samples of various ages at relatively low magnifications. One should note that Everett's (1958) model does not require that the "non-porous" end-member be located near the centre of the solid, although this generally appears to be the case for the radiolaria and diatoms we have observed.

In Fig. 6.24 transmission electron microscopy and nitrogen adsorption observations are combined into a model which will allow a correlation of one type of observation with the other. That is, given the values of specific surface area and specific pore volume for a given assemblage, one should be able to say what an *average* cross-section of the skeleton would look like using transmission electron microscopy and vice versa.

A set of end-member properties (calculations and assumptions in Hurd et al. (in press)) based on Everett's (1958) interconnected rod model and transmission electron microscopy and nitrogen absorption work have been selected. The "non-porous" end-member is assumed to have zero specific pore volume and 1×10^4 cm^2 g^{-1} specific surface area while the porous end-member has about 0.65 cm^3 g^{-1} specific pore volume and 200×10^4 cm^2 g^{-1} specific surface area. Also, "non-porous" core area fraction (core cross-sectional area/total cross-sectional area as seen in a TEM photo) and core radius fraction (core radius/total radius) have been plotted. The core area fraction is equal to the square of the core radius fraction. Thus, if the gas adsorption properties of an assemblage plot near A in Fig. 6.24, one can suggest that the average skeleton will have a very thin porous outer layer; an assemblage having the properties of B should be composed mostly of porous material. Older assemblages appear to have a greater percentage of the "non-porous" skeletal material than the younger ones do.

It is also interesting to consider the ratio of the area of the open pores (internal area) to the external surface, i.e. the porous layer–bulk solution interface. The ratio of internal to external area for porous layers of varying thickness and geometry is as follows:

1. For flat plate, area of pores per unit volume of pores + solid is

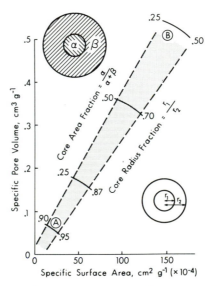

FIG. 6.24 Specific pore volume vs. specific surface area, this time incorporating preliminary data from transmission electron microscopy work. The shaded grey area gives the general range of values for data. As explained in Hurd *et al.* (in press) a set of properties is assumed for the porous and non-porous layers (as seen by TEM) enabling surface area and pore volume estimates for a given skeleton. Either the core area fraction, $\alpha/(\alpha + \beta)$ or the core radius fraction r_1/r_2 can be used to obtain these estimates, which allow the extent of dissolution that a given skeleton has undergone to be determined (figure from Hurd *et al.*, in press).

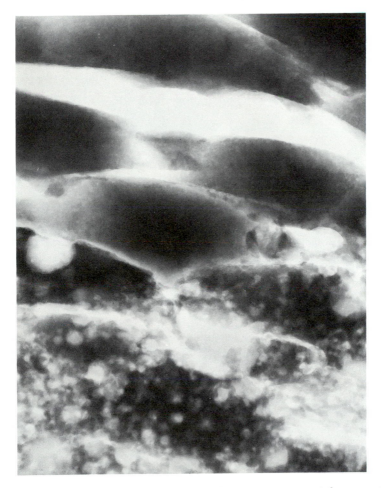

PLATE 6.1. Transmission electron micrograph (128 000 × ; 1 mm ≅ 78 Ångstroms) of a thin section of one of the structural elements of a Recent radiolarian, *Eucyrtidium* sp. The parallel, conchoidal fractures are artifacts of the sectioning process. Note the particulate nature of the less porous phase, wherein many particles about 100 to 150 Ångstroms in diameter appear to be cemented together.

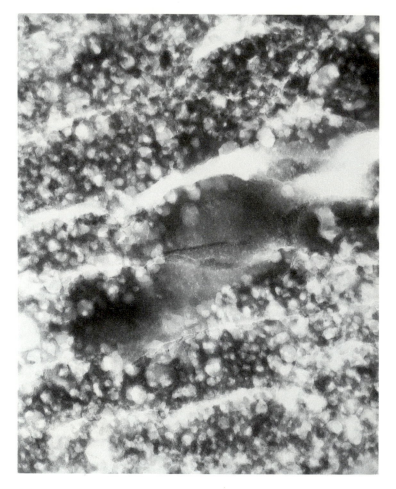

PLATE 6.2. Transmission electron micrograph (128 000 × ; 1 mm ≅ 78 Ångstroms) of a Miocene (about 15 mybp) radiolarian, *Calocycletta virginis*. Parallel, conchoidal fractures are artifacts of the sectioning process. The less porous phase now appears to have more distinct closed pores, but the 100 to 150 Ångstrom diameter blebs still exist although they are less well-defined.

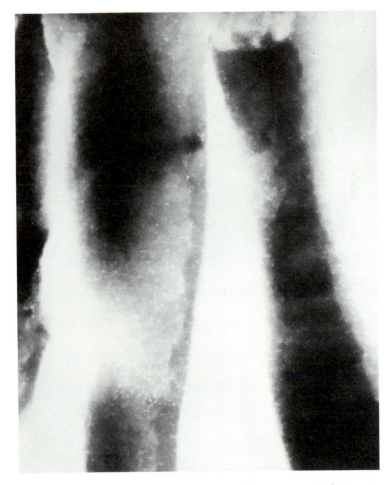

PLATE 6.3. Transmission electron micrograph (128 000 × ; 1 mm ≅ 78 Ångstroms) of an Eocene (about 40 mybp) radiolarian, *Podocyrtis ampla*. Parallel, conchoidal fractures are artifacts of the sectioning process. The particulate nature of the less porous phase is poorly defined but fracturing appears now to occur only layers (centre fragment). The less porous phase also now has relatively large closed pores which are well rounded.

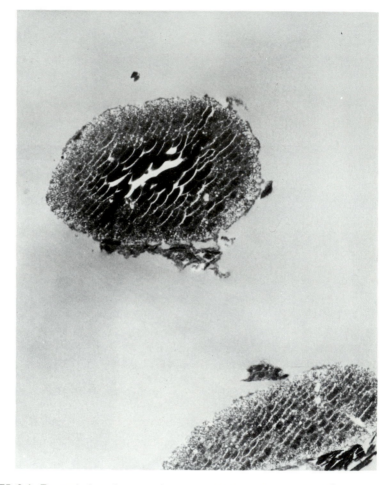

PLATE 6.4. Transmission electron micrograph (9000 × ; 1 mm ≅ 1120 Ångstroms) of a Recent radiolarian, *Eucyrtidium hexasticum*. Parallel, conchoidal fractures are artifacts of the sectioning process. An open-pored outer layer surrounds a less porous or close-pored inner layer.

PLATE 6.5. Transmission electron micrograph (9000 ×; 1 mm ≅ 1120 Ångstroms) of a Miocene (about 15 mybp) radiolarian, *Calocycletta virginis*. Parallel, conchoidal fractures are artifacts of the sectioning process. Note the lesser amounts of outer open-pored skeleton compared with Plate 6.4, which would correspond to lower surface area and pore volume measurements.

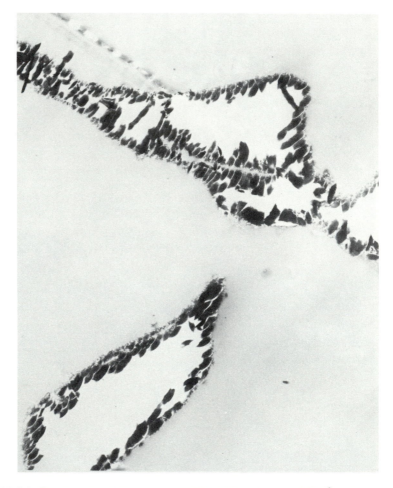

PLATE 6.6. Transmission electron micrograph (9000 × ; 1 mm ≅ 1120 Ångstroms) of an Eocene (about 40 mybp) radiolarian, *Podocyrtis ampla.* Parallel, conchoidal fractures are artifacts of the sectioning process. Note the near absence of a porous outer layer, suggesting very low surface areas and pore volumes.

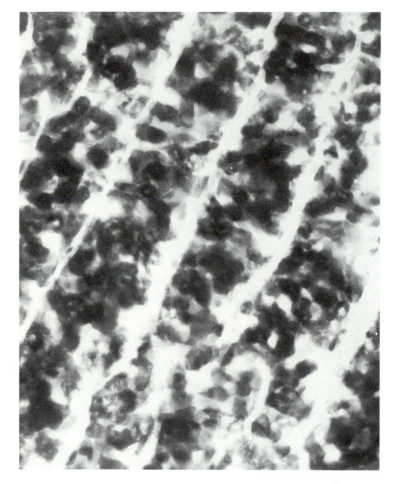

PLATE 6.7. Transmission electron micrograph (128 000 × ; 1 mm ≅ 78 Ångstroms) of a Recent radiolarian, *Phaeodaria castanella,* obtained from sediment traps. Note what appear to be many hollow tubes in this structure.

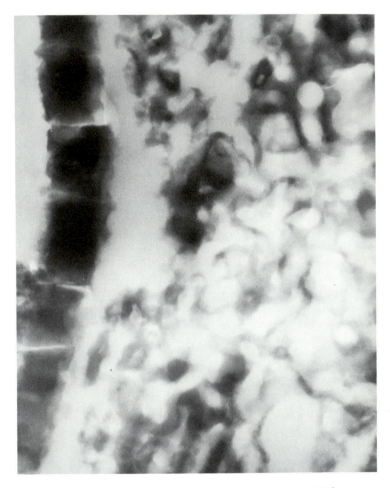

PLATE 6.8. Transmission electron micrograph (128 000 × ; 1 mm ≅ 78 Ångstroms) of a Recent radiolarian, *Phaeodaria challengeron,* obtained from sediment traps. In this case the skeleton appears to be a collection of broken hemispheres and twisted sheets, surrounded by a thicker less porous outer layer.

PLATE 6.9.   Transmission electron micrograph (52 500 × ; 1 mm ≅ 190 Ångstroms) of a Recent radiolarian, *Phaeodaria conchellium capsula,* obtained from sediment traps. This skeleton appears to be composed of at least three morphologically different layers.

PLATE 6.10. Transmission electron micrograph (52 500 × ; 1 mm ≅ 190 Ångstroms) of a Recent species of polycistine radiolarian, *Xiphosphaera* sp., obtained from sediment traps. Note that there is only a very thin outer open-pored layer as compared with Plate 6.4; whether this layer is now forming as a result of dissolution or whether it represents an intermediate step in the skeleton formation is unknown at this time.

200×10^4 cm^2 (volume solid + volume pores)$^{-1}$ = $(200 \times 10^4$ cm^2) $(0.45$ cm^3 + 0.7 cm^3) $\cong 170 \times 10^4$ cm^2 cm^{-3} porous layer; this is the area associated with an area, say, 1 cm^2 × 1 cm thick. So the ratio of internal to external area is: Ratio = $(170 \times 10^4$ cm^2 cm^{-3}) (thickness of layer in cm). A 1 μ thick layer has about 170 times more internal than external area.

2. For a cylindrical rod—total radius R and a non-porous core of radius r then for a unit length of cylinder ignoring the areas of the ends, the porous area, $A_{porous} = \pi(R^2 - r^2)$. In this case the external area equals $2\pi R$; if R equals 1 μ and r is zero then the ratio of internal to external area is (A_{porous}) $(170 \times 10^4$ cm^2 cm^{-3})$(2\pi R)^{-1}$ and is equal to about 85 times more internal than external area, i.e. half the ratio for the plate in (1) above.

   Similar arguments obtain for the case of a sphere, and here the ratio values equal one third if R = 1 micron and r = zero.

3. The gross morphology of radiolarians and diatoms is often best described mathematically as a collection of intersecting cylindrical rods of radius R and distance between adjacent rod centres D (Everett, 1958). Then the ratio of rod area to volume is:

$$\frac{\text{Rod volume}}{\text{Rod Area}} = \frac{7.535R^3 + 9.425R^2(D - 2R)}{3.75R^2 + 18.85R(D - 2R)}$$

Depending on the ratio of D to R and the depth of the porous layer the ratio of maximum internal to external surface area varies between about half to slightly more than one-third the value of the plate in (1) above.

But within a given *assemblage* it is important to note that a *range* of specific surface areas and pore volumes exists which may be a function of inter- and intra-species variability and the state of preservation of a given skeleton (Hurd et al., 1979). The position of one species relative to another on a V vs. A plot may in part explain why that species is better or more poorly preserved than the other. That is, if the dissolution of siliceous material is a two-stage process, wherein a more readily soluble material dissolves first, leaving behind a porous skeleton which also subsequently dissolves, presumably at some slower rate, then the increase in the porosity of the skeleton is a measure of the degree of dissolution which the skeleton has already undergone. The increase in porosity further suggests an increase in the skeleton's friability and thus its ease of destruction. Earlier work on decomposing phytoplankton (Lawson et al., 1978) suggested that the preservation of a given frustule is a function, among other things, of its effective particle size as determined by specific surface area measurements. A greater proportion of porous end-member would yield a smaller effective particle size and result in more rapid disappearance and breakage of the skeleton.

Although most of the polycistine radiolarians and diatoms from sediments we have observed have generally uniform porous layer characteristics, preliminary study of a suborder of radiolarians called phaeodarians (Hurd and Takahashi, research in progress) suggests their ultrastructures are quire different. Plates 6.7, 6.8, and 6.9 show three species of phaeodarians, one of which *Phaeodaria castanella*, appears to be a collection of hollow rods; a second, *Phaeodaria challengeron*, a collection of broken hemispheres; and a third, *Phaeodaria conchellium capsula*, containing at least three distinct layers, all having different morphologies. All specimens were collected using the PARFLUX Mark II sediment traps. A more extensive description of these and other phaeodarians is in preparation.

In sharp contrast to the phaeodarians, preliminary results involving the study of the skeletal ultrastructure of *polycistine* radiolaria from the PARFLUX Mark II sediments traps, using both transmission electron microscopy and methyl red due adsorption, have shown these skeletons to have few open pores. Several genera which have been commonly observed as porous in the sediments (i.e. having many open pores) are essentially either non-porous or have mostly closed pores when they are found in sediment traps (Plate 6.10). One can tentatively interpret these findings as being supportive of the model of biogenic silica as a two-component system which dissolves in several stages. The following arguments based on physical property observations form the basis of this model.

Figure 6.25 shows a plot of assemblage specific pore volume versus the *mean* skeletal densities of a number of species within the assemblage as determined by "heavy" liquids (methods and materials described in Hurd and Theyer, 1977). There are at least two choices of models when dealing with porous solids: (i) the solid is a single component system and all pores which *do* occur are open to the outside; (ii) the solid is a two or more component system or a single component system having closed pores filled with yet another component(s). In the first case, if the pores are small, i.e. less than 20–40 Å in radius, they would spontaneously fill with an adsorbent whose saturation vapour pressure was greater than about 70%. For example, since the water vapour pressure in the air at the Manoa campus of the University of Hawaii, where the studies were conducted, is in the range 63–73% in the course of a year, one would expect that the siliceous assemblages would have a large percentage of their external pores filled with water if they were stored at room temperature and open to the atmosphere. The solid line in Fig. 6.25 estimates the skeletal density of an externally porous silica sample, whose solid density is 2.26 g cm^{-3} and which is being filled with water of density 1.0 g cm^{-3}. The bar-dot line shows the same solid being filled with carbon tetrachloride ($\sigma = 1.6$ g cm^{-3}), the less dense component of the "heavy" liquids. Only at very high pore volumes do any of the data begin to approach even the water–silica line. At low pore volumes (remembering that these are pores open to the outside being measured), the density values cluster between 1.9 and 2.0 g cm^{-3}. If the two components of the skeleton were water and silica having the above densities (1.0, 2.26 g cm^{-3} respectively), the weight per cent water range would be 8–15. This range corresponds well with the water contents of "non-porous" samples studies earlier by Hurd and Theyer (1977) sample numbers 34, 36, and 28 and the sponge spicules all have surface areas less than 10 m^2 g^{-1} and therefore pore volumes less than 0.02 cm^3 g^{-1} (see also Hurd *et al.* (in

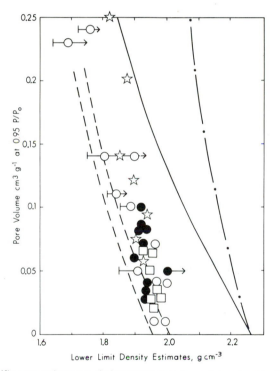

FIG. 6.25   Specific pore volume vs. skeletal bulk density as determined by "heavy" liquids. Note that at near zero pore volume the densities still cluster around 1.9 to 2.0 g cm^{-3} rather than 2.25 g cm^{-3}, suggesting the possibility that a two-component system exists. The solid line = simple mixing of 2.25 g cm^3 non-porous silica and 1.0 g cm^{-3} density water; pair of broken lines = mixing of density 1.95–2.0 skeletal material with water, and the dot-dash line = mixing carbon tetrachloride (one of the heavy liquids) with 2.25 g cm^{-3} silica. The trend in the data appears to go from the lower skeletal density to the higher, suggesting that as the skeleton dissolves the lower density two-component mixture with closed pores opens to leave the higher density material behind.

press)). That the sample contains substantial amounts of water yet has little external porosity suggests that the water exists in the closed pores within the solid. Plate 6.3 shows a high resolution TEM photograph of portions of an Eocene age radiolarian skeleton from sample 28 showing the existence of small widely distributed pores in the solid, thus partially substantiating the model.

Recent TEM work dealing with freshly fixed diatoms (Drum and Pankratz, 1964) and sponges (Simpson and Vaccarro, 1974) neither of which had been acid-cleaned, showed that these organisms precipitate a porous silica structure which is later infilled to some extent, and perhaps traps water during the infilling process. It can now be suggested that after the death of the organism when the process of skeletal dissolution occurs more rapidly, that a porous structure is again produced, this time by the preferential dissolution of the infilled material and the opening of pores previously sealed by the infilling process. If during the last 10 million years or so, the relative number of small closed pores increased, then such skeletons would be more susceptible to porosity formation during the dissolution process. The greater proportion of such pores might also partially explain why younger material is more porous than older material (Fig. 6.21) even though the latter has had more time to react with its surroundings. One might argue that the older material was porous and later infilled. If this is so then the infilling must have occured using primarily 2.25–2.30 density silica and not 2.6 density silica (quartz), since the preliminary density vs. water content relationships found do not support the latter, even though the solubility data does. It is probable that careful oxygen isotope analysis of the siliceous material would show whether or not extensive infilling had occurred in these older samples and had thus rendered them less porous by that process.

In earlier work (Hurd, 1973; Hurd and Theyer, 1975, 1977) radiolarian and diatom skeletons were cleaned from sediments using hot, dilute HCl, sieving and ultrasonic vibration in an attempt to dissolve and break loose clays and other materials which were adhering to and within the skeletons. This approach was felt to be generally successful in that a microscopically clean sample usually resulted. Those samples with obvious amounts of contaminants were now knowingly subjected to the physical properties survey. Recently the adsorption of methyl red dye dissolved in benzene onto the skeletons of diatoms and radiolarians has been studied in order to provide a method for the estimation of skeletal porosities of individual specimens as well as assemblages. If such a method were to work, a great deal of time could be saved when choosing samples for further TEM study.

The adsorption of the acid-base indicator methyl from solutions of benzene or carbon tetrachloride has been used to estimate the specific surface area of silica gels for more than 25 years (see Hurd et al., in press). The use of this material for determining the specific surface area of biogenically precipitated silica has been reinvestigated and it was found that, although this method is operationally cumbersome, good agreement exists between methyl red-determined surface area values and those determined by nitrogen adsorption. Figure 6.26 shows a portion of the data giving the relationship between the optical absorbance of a solution of methyl red in benzene at 480 mμ and the *total* area of the sample in m² (determined by nitrogen adsorption) of four siliceous assemblages. This experiment shows that the silica skeletons adsorb methyl red in a manner both consistent with, and proportional to, the nitrogen adsorption area. This also means that by observing the colour *intensity* of the dyed skeletons one can obtain a semi-quantitative measure of the surface area, and therefore the porosity of even a single skeleton if desired (Adams and Voge, 1957). More porous material will stain more deeply because there is more surface for the dye to adsorb onto (and because of the porous nature of the material more layers of biogenic silica with dye adsorbed on them will be looked through), as well as more pores for the dye solution to become trapped in during the drying and slide preparation process.

A number of samples from the sediments showed heterogeneity in the form of blotchiness on the skeletons and small bundles of deeply stained material within the skeletal framework. The blotchiness and small bundles could now

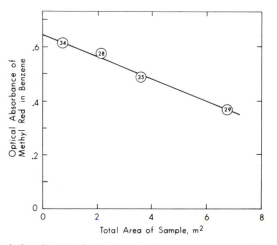

FIG. 6.26   Optical absorbance of methyl red in benzene vs. the total area of a given silica sample (as determined by nitrogen adsorption). The linear relationship between dye adsorption onto the silica surface (as evidenced by the decrease in absorbance of the dye solution) and the sample area suggests that dye adsorption onto various siliceous skeletons is an effective method for determining whether high surface areas and therefore high porosities exist (Figs 6.25 and 6.26 from Hurd *et al.*, in press).

be readily distinguished using light microscopy because of the colour intensity differences between the contaminants and the biogenic silica. Certain contaminants also appeared more orange or yellow than red. This in turn led to the investigation of a series of dyes called the Hammett indicators (Hammett, 1935; Tanaba, 1970 and references therein).

Briefly, if an acid-base indicator is adsorbed to the surface of a solid and the solid is sufficiently more acidic or basic than the indicator, then a proton will be donated to, or removed from, the indicator by the solid and the colour of the indicator will change. By noting the colour of a number of varying acid-strength indicators such as the Hammett series, we can determine the relative acidity or basicicity of a variety of solids (Tanabe, 1970). Furthermore, since clays and combinations of alumina and alkaline earths are more acidic than silica alone, one may also be able to determine whether these materials have been incorporated to some extent onto the surface of biogenically precipitated silica in sediments.

The adsorption of the Hammett indicators onto biogenic silica and aluminosilicates will be considered in more detail in future work. For the present, these indicators are being used primarily to denote contamination of biogenic silica by other minerals, and to give a relative estimate of surface porosity in radiolarian assemblages of various ages.

Contamination by other minerals shows up as blotchiness (in the case of methyl red) or by the aluminosilicates being one colour and the biogenic silica being another because of differences in surface acidity. Thus methyl yellow (p-Dimethylaminoazobenzene) is yellow on the biogenic silica surface and red on the aluminosilicate surface; benzeneazodipheylamine is yellow on the biogenic silica surface and generally purple on the aluminosilicate surface.

6.5 USE OF X-RAY DIFFRACTION FOR ANALYSIS OF BIOGENIC SILICA AND AMORPHOUS SILICAS

The X-ray diffraction and thermal inversion behaviour of biogenic silica, silica gel, and vitreous silica, all before and after various heat treatments is of considerable interest. Such behaviour may give clues to the varying degrees of order and disorder which seem to exist in these different substances. Consider first some of these properties in the light of their empirical application as methods for biogenic silica in sediments by oceanographers and the compare work done in other fields and other approaches toward understanding biogenic silica from this viewpoint.

As originally described by Goldberg (1958), biogenic silica, which exhibits a relatively poor diffraction pattern (Figs. 6.27, 6.28, 6.29), is converted to low-cristobalite to some unknown but reproducible degree by heating a carbonate-free, finely ground sample at 900°C for 4 h. The percentage biogenic silica is determined by known amounts of this same substance being added to each sample and the original amount of material present determined by solving a set of standard addition equations. This method appears to be a reasonable empirical approach (i) if samples of un-contaminated biogenic silica are removed from another subsample of the *same* sediment and used for the standard addition process; (ii) if the process of separation and decontamination does not alter the assemblage present or the manner in which a given assemblage reacts to the temperature treatment; and (iii) if the percentage biogenic silica in the sediments is not less than about 5%.

Calvert (1966) modified the above method by using alumina (alpha-Al_2O_3) as an internal standard and increasing the heating temperature to 1000°C; he

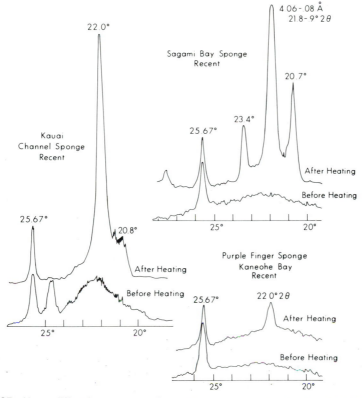

FIG. 6.27 X-ray diffraction patterns of Recent sponge spicules. Peak at 25.67°2θ is an internal standard, alpha-alumina (figure from Hurd and Theyer, 1977).

noted that differences exist between standard curves prepared for diatoms, radiolarians, and sponge spicules but does not give the degree of error involved. Again, the method appears to be a reasonable empirical approach (i) if the standard curves for the three different siliceous materials studied apply uniformly to all siliceous sediments; and (ii) if there is no interference with, or production of low-cristobalite by, other materials in the sediment. Recent studies, however, suggest that neither of these criteria have been successfully met (Eisma and Van Der Gaast, 1971; Jones and Segnit, 1971; Greenwood, 1973; Keene, 1973; Hurd and Theyer, 1977).

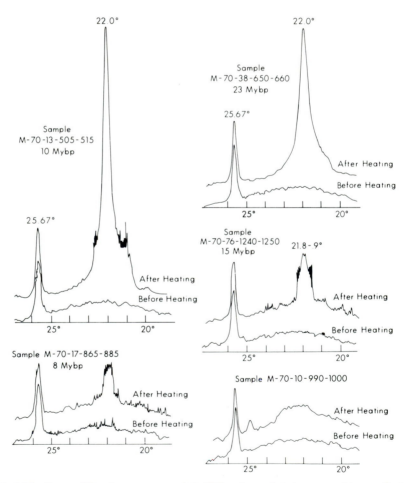

FIG. 6.28   X-ray diffraction patterns of 8–17.5 mybp radiolarian assemblages. Peak at 25.67°2θ in an internal standard, alpha-alumina (figure from Hurd and Theyer, 1977).

With respect to both of the above studies (Goldberg, 1958; Calvert, 1966) several things are curious. In neither case was the amount of low cristobalite produced by heating a given amount of biogenic silica ever compared with a pure low-cristobalite sample also mixed with an internal standard. As a result, it is unsure to what extent any given sample was ever "converted" to low cristobalite. This is important from structural considerations as will be discussed in greater detail later, but for the moment it is interesting simply to observe in Figs 6.27, 6.28, and 6.29 the variability with respect to "conversion" to low cristobalite of a number of different HCl-cleaned

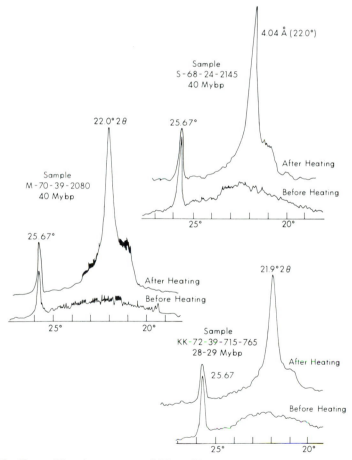

FIG. 6.29   X-ray diffraction patterns of 28 to 40 mybp radiolarian assemblages. Peak at 25.67°2θ is an internal standard, alpha-alumina (figure from Hurd and Theyer, 1977).

biogenic silica samples which were X-rayed before and after heating to 1000°C for 24 h (Hurd and Theyer, 1977). The internal standard used was alumina and sample preparation the same as Calvert (1966). Also shown are the X-ray diffraction patterns of low cristobalite and opal-CT or porcellanite/lussatite (Calvert, in press) for reference (Fig. 6.30). As is readily apparent, the degree of "conversion" goes from practically nil to some unknown but higher amount. It is further interesting to note that the products either produced or revealed by the heating method are by no means uniform in composition with respect to the number of peaks which appear or disappear.

In this respect the experimental work by Verduch (1955), Hill and Roy (1958), and Foster *et al.* (1966) provides a useful set of criteria to consider. In one set of experiments, Verduch heated 99.9% pure amorphous silica to temperatures between about 950°C and 1100°C for up to 60 h. His results are shown in Figure 6.31. The vertical axis is labelled "apparent cristobalite content (%)" because the heated samples, which Verduch felt contained unaltered amorphous silica, partly altered amorphous silica/disordered cristobalite, and well-crystallized cristobalite, were compared with a standard curve containing only amorphous silica and well-crystallized cristobalite as end-members. It is important to note, however, that even when samples were heated twice as long as the X-ray diffraction method suggests, approximately only one-third of the sample had been altered, after which essentially no further change took place.

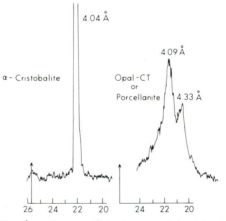

FIG. 6.30 X-ray diffraction patterns of both pure alpha-cristobalite, formed at high temperatures, and opal-CT, a mixture of polytypes of cristobalite and tridymite, in this case with biogenic silica as a precursor. Arrows on the left of each graph locate the position of the internal standard, alpha-alumina at 25.67°2θ.

In the Hill and Roy work, various samples of nearly pure silica were heated to various temperatures for various times and the *inversion temperature* of the cristobalite produced was measured. Their data is shown in Fig. 6.32. Here again, if one regards the heating cycle inversion temperature of 267°C as the upper limit for pure cristobalite, then in no case was the beginning material completely converted until the lower stability field temperature (about 1470°C) had been reached. Further, the authors state:

> ... thus pure silica gel heated for progressively longer times yields corresponding cristobalite phases with increasing inversion temperatures from ca. 130°C to 267°C. Silica glass (very low in impurities) can also be crystallized at 1600°C to give the cristobalite phases with the highest inversion temperatures, near 267°C, or it can be made to yield cristobalites with lower inversion temperatures, depending on the heat treatment.

These authors also rank what are commonly called X-ray amorphous silica in the following manner with respect to ease of conversion to cristobalite by heating: (i) silica gel, freshly precipitated by the hydrolysis of ethyl orthosilicate using reagent grade ammonia in platinum or silica vessels, is the *least* ordered; (ii) silica glass is next; (iii) silicic acid, which they suggest is essentially an aged, dried gel, is best ordered, although their data as arranged in their Fig. 6.32 suggests this order may change depending on heating temperature.

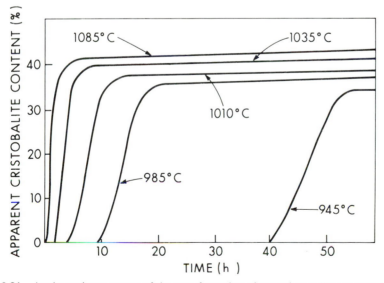

FIG. 6.31.   Isothermal rate curves of the transformation of amorphous silica to cristobalite (figure from Verduch, 1958).

Foster *et al.* (1966) considered the position of the cristobalite d(101) peak, the intensity of that peak relative to an internal standard (BaF_2), the ratio of the 3.15 Å to the 4.05 Å cristobalite peaks, the volume change associated with the inversion of low to high cristobalite and the slope of the expansion versus temperature curve during the inversion process. All of these parameters have been suggested as being indicative of crystalline disorder. In agreement with Verduch (1955), Foster *et al.* found a strong correlation between d(101) spacing and peak height or ratio to internal standard. Foster *et al.* also found strong correlations between these same parameters and both the total volume change and rate of change at inversion. They also note that "crystalline disorder" (their term) did not alter variations in thermal expansion properties; if the disorder were of a vitreous nature this would be reasonable in that the thermal expansion properties of vitreous silica are about 10 times less than cristobalite, tridymite, or quartz in this temperature range.

The author has been unable to find any discussion in the literature regarding the change in refractive index when biogenic silica sample is heated

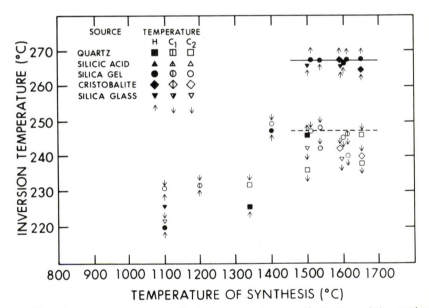

FIG. 6.32   Diagram illustrating the effect of heat-treatment and structure of the starting material on the inversion temperature of cristobalite. Impurity effect are completely excluded both by the purity of the starting material and by the fact that the same starting material is affected differently by different heat treatments. Closed symbols with arrow pointing upwards = temperatures obtained on the heating cycle; open symbols = temperatures on the cooling cycle; symbols as listed under column C_1 indicate more than one peak on the cooling cycle (figure from Hill and Roy, 1958).

to 1000°C for 24 h except for the work of Hurd and Theyer (1977). Since the initial refractive index of a given biogenic silica sample is a function of both the silica molecular arrangement and ratio of water to silica in the skeleton, and since the sample both loses water, sinters, and decreases its specific surface area when heated, if all porosity were eliminated and only low cristobalite were produced, the resultant refractive index should then be 1.485. This number was not reached by any of the samples tested. It might immediately be said that some sample porosity remained and the resultant indices measured were a combination of both air and silica. Figure 6.33 shows change in relative intensity of low cristobalite to alumina versus refractive index of the *heated* samples. Figure 6.34 shows the same relative intensity change versus the change in refractive index (heated minus unheated sample values). The amount of variability is quite large but it is difficult to completely deny that both those samples with low refractive index values after heating and those having a small change in refractive index generally have smaller intensity changes associated with them. This may further suggest that for some reason the molecular arrangement of silica in these samples is more resistant to thermal alteration than those with larger changes.

In retrospect, it may be wondered why any changes in low-cristobalite intensities occur at all, since the actual stability field of cristobalite is between 1470°C–1723°C, although metastable cristobalite can and does exist at all lower temperatures. This is discussed at some length by Sosman (1965), and

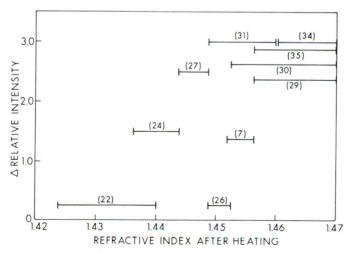

FIG. 6.33 Change in the relative intensity of low cristobalite to alpha-alumina vs. the refraction index; both measurements on heated samples (figure from Hurd and Theyer, 1977).

references therein as well as Rieck and Stevels (1951), Muroya and Kondo (1970), and Strelko *et al.* (1974). In all probability, the presence of small amounts of various alkali metal and alkaline earth oxides which act as fluxes are partially responsible for those changes which do occur since high purity silica gel does not readily convert to low cristobalite when heated to only 1000°C.

In an attempt to overcome the foregoing difficulties involving heat treatment of the sample to produce low cristobalite, Eisma and Van der Gaast (1971) introduced the notion of using the height of the amorphous silica "bulge" at 26° 2θ as well as the area under the curve between 17° 2θ and 40° 2θ using Co K alpha radiation and alumina as an internal standard. The amount of error above 10 wt. % biogenic silica is given as ±3% of the value found but below 10 wt. % the error rapidly approaches the wt. % biogenic silica present. In addition, the authors mention that sediments having large amounts of illite, kaolinite, mica, and montmorillonite interfere, an observation also noted by Zemmels and Cook (1973). Eisma and Van Der Gaast suggest that standard addition of biogenic silica to the sample analogous to the method of Goldberg (1958) still allows their method to be useful, but now with a larger error of at least ±6%; however, neither their correction procedure nor method of applying it to sediments in general was given in any detail. The method appears most applicable to sediments containing more than 10 wt. % biogenic silica and having only calcite, quartz, feldspars, and sparing amounts of clays present.

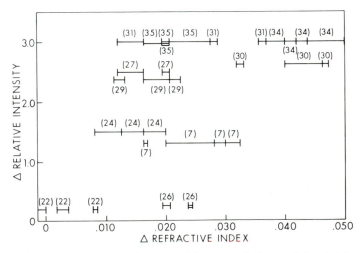

FIG. 6.34 Change in the relative intensity of low crostobalite to alpha-alumina vs. the change in refractive index of each species in several samples; both measurements on heated samples (figure from Hurd and Theyer, 1977).

It would seem at this point that a promising line of research might be one of using the "conversion" method of Goldberg (1958) and Calvert (1966) not for quantitative analysis of biogenic silica in sediments but rather as a tool for studying the types and organization of silica laid down by various organisms. That is, a sample which has the silica tetrahedra so arranged that they very nearly are in the low-cristobalite arrangement will require less energy to enhance this structure than say, vitreous silica. The intriguing point here is that heating a given sample to 1000°C for about 24 h in no way produces a uniform effect on the assemblage *per se*. As evidenced especially by refractive index measurements, there are differences from one species to the next, within a given species, and even from one portion of the skeleton to another.

Gregg *et al.* (1977) have suggested that small amounts of strained quartz occur in the skeletons of the several species of freshwater diatoms they studied. Although it is difficult to interpret their "index of crystallinity" (wherein chert has a higher value than quartz), at least their general approach of detailed peak scanning should be considered. It would also be interesting to see if the rate of nucleation of silica gel as controlled by temperature, pH, and ionic strength, would have any effect on the production of quartz or cristobalite. One could imagine that statistically, all polymorphs stable at one atmosphere might be produced to some extent. The proportions of each would have to be a function of the free energy of activation of the solid relative to the monomer (Hurd and Theyer, 1975) and the various possible successful orientations of tetrahedra that result in a given polymorph.

6.5.1   Questions Relative to the X-ray Diffraction Section

1. When a given biogenic silica sample is heated to 1000°C for about 24 h, what is the percentage cristobalite produced relative to a pure cristobalite standard (as well as an internal standard)?
2. What is the relationship between peak height to peak width ratio as a method for estimating crystallite size as compared with small-angle X-ray scattering to measure ultimate particle size?
3. Do those portions of biogenic silica which do not change their refractive index and X-ray pattern when heated have a structure similar to vitreous silica?
4. Given such energy diagrams as relatively free energy of activation between monomer and various silica polymorphs, is it possible to predict at a given temperature the various proportions of each polymorph produced and in turn relate these predictions to the available and necessary number of successful orientations of tetrahedra for each polymorph?

6.6   INFRA-RED SPECTROSCOPY

In a manner similar to the previous discussions, several empirical methods for the analysis of biogenic silica in sediments are discussed here before considering infra-red spectroscopy as a tool for better understanding the interactions of silica and water.

Chester and Elderfield (1968) developed methods for opal and quartz determinations in carbonate-free sediments wherein a portion of the very finely ground sediment (all of the sample must be less than 2 μm) is pressed into a KBr pellet and the absorption of the 800 cm^{-1} peak is measured. The quartz doublet at 780 and 799 cm^{-1} interferes when quartz is present in amounts greater than 5% on a carbonate-free basis, but this interference can be largely eliminated by means of a balancing disk technique wherein the amount of quartz present is estimated using the height of the 695 cm^{-1} peak and a KBr pellet is prepared having that concentration of quartz, and placed in the second beam of a twin beam spectrophotometer. Kaolinite also interferes but its interference can be largely eliminated by hand grinding the sample for 20 min in a corundum mortar and pestle. When the authors compared this method with that of Calvert (1966) using 15 of the same samples, the agreement was within about 3–5% absolute, which is quite good. Further, the precision of the method can also be high, depending on the composition of a given sample. One of basic difficulties of the method, however, is that all silica polymorphs absorb at about 800 cm^{-1}, which represents an Si–O–Si stretching mode (Lippincott et al., 1958; see also Fig. 6.35), so that a mixture of quartz, opal-CT, and biogenic silica (or opal A), such as those found in DSDP cores might be extremely difficult to separate analytically. In addition, in a recent talk given at the 25th International Mineralogical Congress, at Sydney, Australia, Jones and Flörke discussed the band width and intensity of the 800 cm^{-1} peak for a large variety of silica polymorphs. In general, X-ray amorphous materials had much greater band widths and were less intense than more ordered structures; samples of noncrystalline material given more well-ordered patterns as they were exposed to increasingly greater temperatures. Since both the X-ray and infra-red methods respond to well-ordered structures much better than poorly ordered ones, this may be the basis of their agreement; and, if all biogenic silica had the same amount of order or disorder, then either or both would be acceptable. But since biogenic silica apparently is not uniform, as suggested by the "conversion" or low-cristobalite data, then it is not sure how much of the total silica present is actually being measured by either method.

Bogdanov et al. (1974) describe a method wherein the sample is mixed with a given amount of potassium rhodanate (KSGN) and made into a paste by adding a drop of petrolatum. The potassium rhodanate absorption peak at 750 cm^{-1} serves as an internal standard analogous to the alumina and/or BeO in the X-ray method; this absorption peak is then compared with the biogenic silica peak at 800 cm^{-1} and the quartz peak at 695 cm^{-1} to determine both substances. They mention that both biogenic silica and quartz are most accurately analysed when samples contain more than 10 % of either substance; quartz interferes with the analysis of biogenic silica, but not

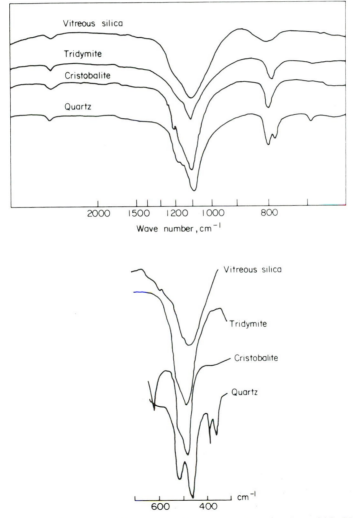

FIG. 6.35 Infra-red absorption by various silica polymorphs from about 300–3000 cm^{-1} (figure after data in Lippencott *et al.*, 1958).

vice versa; it is also noted that volcanic glass, certain montmorillonites and zeolites interfere with the analysis, suggesting that much of the northwestern, equatorial and south-eastern Pacific sediments would be difficult to analyse by this method.

Kamatani (1971, 1974) has observed the infra-red spectrum of several species of diatoms over the range 400 to 4000 cm^{-1} as a function of heating the samples to 700°C. Dehydration of the samples was reversible up to 500°C and the loss of surface silanols was followed by observing the disappearance of the absorption peak at 950 cm^{-1}. Lowenstam (1971) observed that certain radulat sites of molluscs and gastropods also contain large amounts of amorphous silica which has an infra-red spectrum similar to a reference opal.

Langer and Flörke (1974) in their thorough study of the near infra-red absorption spectra (4000 to 9000 cm^{-1}) of H_2O and OH in various types of opals arrived at the conclusion that molecular water and hydroxyl groups of several types exist:

1. Molecular water consists of two types, one (type A) which is suggested as being "single, almost non-hydrogen bonded molecules in small cages (diameter about 0.35 nm) of the SiO_2 matrix"; the other (type B) being "liquid-like, hydrogen bonded water as a film on inner surfaces."
2. Water in the form of SiOH-groups of two types, one of which has stronger hydrogen bonds than the other.

In general, molecular water accounts for 60–90% of the total water as determined both chemically and by weight loss on ignition. Surprisingly, in one of their samples studied, about 40% of the molecular water was suggested as being type A. This type of water *increases* both the density and refractive index of a given sample upon addition and thus has nearly the opposite effect of type B molecular water with respect to changing these properties. For example, a 5 wt. per cent addition of type A molecular water to the hypothetical silica polymorph (n = 1.465, d = 2.21 g cm^{-3}) increases the refractive index and density to 1.478 an 2.32 g cm^{-3} respectively. These values were obtained by using equations (17) and (19) and assuming the molar volume of water to be equal to zero, since there would be no net volume change when the water was added. As can be seen by plotting this data point on Figs 6.12, 6.17, and 6.18 the addition of approximately equal amounts of type A and B molecular water might well show no net water addition effect in Fig. 6.3, but strong vertical displacement of the data point in Figs 6.17 and 6.18. Evidence for the existence of such water in obsidian and perlite appears to have been observed by Ross and Smith (1955) as shown in their Fig. 1.

Figure 6.36 shows a simplified version of one of Langer and Flörke's figures with additional information regarding types of stretching and bending frequencies included. The upper curve represents the actual absorption spectrum and the lower curve slows the mathematical separation into Gaussian distributions of various overlapping curves. Molecular water $(v_2 + v_3)$ combinations appear at about 5200 cm^{-1} and silanol group $(v_{OH} \times \delta_{SiOH})$ combinations are in the region about 4500 cm^{-1}. Weaker molecular water fundamentals $(v_1 + v_3)$ also exist at about 7000 cm^{-1} but were not used because of the better resolution of the $v_2 + v_3$ pair at about

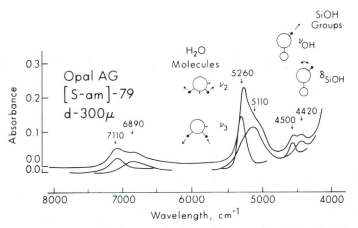

FIG. 6.36 Infra-red absorption by an amorphous silica sample of 300 μM thickness from about 4000–8000 cm^{-1} (figure drafted from a figure and table in Langer and Flörke, 1974).

5200 cm^{-1}. (For a more thorough discussion and interpretation see original article.) Thus far there do not appear to be any articles dealing with a detailed study of the near-infra-red absorption of water and silanol (SiOH) groups in biogenically precipitated silica such as Langer and Flörke's. Yet such studies, if combined with transmission electron microscopy, would be particularly useful with respect to water removal from biogenic silica for isotope studies. Although present methods usually involve flash heating, that molecular water appears to exist in discrete closed pores suggests grinding under vacuum may be an approach which would be less liable to alter isotope ratios. Further, if the pre-existing amount of type A silanol groups could be determined for a given species of diatom or radiolarian then it might be possible to correlate these amounts with its rate of dissolution; the amount of type A silanols should be proportional to particle areas as well as giving the extent of partial hydrolysis of the silica by the water.

REFERENCES

Aristov, B. G., Karnaukhov, A. P., and Kiselev, A. V. (1962). *Russian Journal of Physical Chemistry*, **36(10)**, 1159.

Adams, C. and Voge, H. (1957). *J. Phys. Chem.*, **61**, 722.

Batsanov, S. S. (1961). "Refractometry and Chemical Structure". Consultants Bureau, New York, New York.

Baumann, H. (1965). *Beitrage zur Silikose-Forschung* **85**, 1.

Bogdanov, Yu. A., Levitan, M. A., and Plyusnina, I. T. (1974). *Oceanology*, **14(5)**, 756.

Brewer, P. and Hao, W. M. (1979). *In* "Chemical Modeling in Aqueous Systems" (Jenne, E. A., ed.). ACS Symposium Series #93, Am. Chem. Soc., Washington, DC.

Burckle, L. (1979). *In* "Introduction to Marine Micropaleontology" (Haq, B. U. and Boersma, A., eds.). Elsevier, New York.

Calvert, S. E. (1966). *Geol. Soc. Am. Bull.*, **77**, 569.

Calvert, S. E. (1974). *In* "Pelagic Sediments on Land and Under the Sea". Spec. Pub. Is. Int. Ass. Sediment, **1**, 273.

Carman, P. C. and Raal, F. (1951). *Proc. Royal Soc.*, **209A**, 59.

Chester, R. and Elderfield, H. (1968). *Geochim. et Cosmochim. Acts*, **32**, 1128.

Daniels, F. and Alberty, R. A. (1961). "Physical Chemistry". 2nd edition, John Wiley, New York.

Dawson, P. A. (1973). *J. of Phycology*, **9**, 353.

Demaster, D. (1979). "The Marine Budgets of Silica and ^{32}Si". Unpublished PhD thesis, Yale University, Department of Geology and Geophysics, New Haven, Conn., USA.

Drum, R. W. and Pankratz, H. S. (1964). *J. Ultrastructural Res.*, **10**, 217.

Everett, D. H. (1958). *In* "The Structure and Properties of Porous Materials" (Everett, D. H. and Stone, F. S., eds.). Butterworths Scientific Publications, London.

Eisma, D. and Van Der Gaast, S. J. (1971). *Neth. J. Sea Res.*, **5(3)**, 382.

Foster, P. K., Hughes, I. R., and MacKenzie, I. J. D. (1966). *New Zeal. J. Sci.*, **9**, 249.

Fournier, R. O. (1973). *In* "Proceedings of Symposium on Hydrogeochemistry and Biochemistry" Vol. I—Hydrochemistry. US Geol. Survey, Clarke Co., Washington DC.

Fournier, R. O. and Rowe, J. J. (1972). *Am. Mineral.*, **47**, 897.

Frank-Kamenetskii, D. (1969). "Diffusion and Heat Transfer in Chemical Kinetics". Plenum Press, New York.

Frondel, D. (1962). "The Systems of Mineralogy", Vol. III. John Wiley, New York.

Goldberg, E. D. (1958). *J. Mar. Res.*, **17**, 178.

Greenburg, S. A. (1957). *J. Phys. Chem.*, **61**, 960.

Greenwood, R. (1973). *J. Sed. Petrol.*, **43**, 700.

Gregg, J., Goldstein, S., and Walters, L. (1977). *J. Sed. Pet.*, **4**, 1623.

Gregg, S. and Sing, H. S. W. (1967). "Adsorption, Surface Area and Porosity". Academic Press, New York and London.

Griffin, J. (1980). "The Effect of Pressure on the Solubility of 11 Silicates in Seawater at 2°C, pH 8". Unpublished Master's Thesis. Univ. of Hawaii, Department of Geology. Honolulu, Hawaii, USA.

Grill, E. V. and Richards, F. A. (1964). *J. Mar. Res.*, **22**, 51.

Hammett, L. (1935). *Chem. Rev.*, **16**, 67.

Harder, H. (1971). *Mineral. Soc. Japan Spec. Paper*, **1**, 106.

Harper, H. and Knoll, A. (1975). *Geology*, **175**.

Hecky, R. E., Mopper, K., Kilham, P., and Degers, E. (1973). *Mar. Biol.*, **19**, 323.

Hendricks, S. B. and Jefferson, M. E. (1938). *Am. Mineralogist*, **23**, 863.

Hill, V. G. and Roy, R. (1958). *J. Am. Ceram. Soc.*, **41(2)**, 532.

Hurd, D. C. (1972). *Earth Planet. Sci. Letts.*, **15**, 411.

Hurd, D. C. (1973). *Geochim et Cosmochim. Acta*, **37**, 2257.

Hurd, D. C. and Theyer, F. (1977). *Amer. J. Sci.*, **277**, 1168.

Hurd, D. C. and Theyer, F. (1975). Analytical Methods in Chem. Ocean. (Gibb, T. R. P., Jr., ed.). A.C.S. Adv. Chem. Series #147. Am. Chem. Soc., Washington DC.

Hurd, D. C., Wenkam, C., Pankratz, S. and Fugate, J. (1979). *Science*, **203**, 1340.

Hurd, D. C., Pankratz, H. S., Asper, V., Fugate, J. and Morrow, H. (in press). *Amer. J. Sci.*

Hurd, D. C. and Takahashi, K. (Research in progress).

Iler, R. K. (1955). "The Colloid Chemistry of Silica and Silicates". Cornell Univ. Press, Ithaca, N.Y.

Iler, R. K. (1979). "The Chemistry of Silica-Solubility, Polymerization, Colloid and Surface Properties and Biochemistry". Wiley-Interscience, New York.

Jones, M. M. and Pytkowicz, R. M. (1973). *Bulletin de la Societe Royale des Science de Liege*, **42**, 118.

Jones, J. B. and Segnit, E. R. (1971). *Geol. Soc. Australia J.*, **18(1)**, 57.

Kamatani, A. (1971). *Mar. Biol.*, **8**, 89.

Keene, J. B. (1973). *In* "Initial Reports of the Deep Sea Drilling Project", Vol. 32 (Larson, R. L. *et al.*, eds.). US Government Printing Office, 429.

King, K. (1975). *Micropaleontology*, **21(2)**, 215.

Kiselev, A. (1958). *In* "The Structure and Properties of Porous Materials" (Everett, D. and Stone, F., eds.). Butterworths, London.

Kling, S. (1978). *In* "Introduction to Marine Micropaleontology" (Hag, B. and Boersma, A., eds.). Elsevier, New York.

Kokta, J. (1931). *Rozpravy Ceske Akad.*, *Ser. II*, **40(21)**.

Kokta, J. (1931). *Mineralog. Abs.*, **4**, 517.

Krauskopf, K. B. (1956). *Geochim. Cosmochim. Acta*, **10**, 1.

Krauskopf, K. B. (1959). *Soc. Econ. Paleontologists Mineralogists Spec. Pub.*, **7**, 4.

Langer, K. and Flörke, O. W. (1974). *Fortschritte der Mineralogie*, **51(1)**, 17.

Lawson, D., Hurd, D. C., and Pankratz, H. S. (1978). *Am. J. Sci.*, **278**, 1373.

Lerman, A. and Lal, D. (1977). *Am. J. Sci.*, **277**, 238.

Levich, G. V. (1962). "Physico-Chemical Hydrodynamics". Prentice-Hall, Inc., Englewood Cliffs, N.J.

Lewin, J. C. (1961). *Geochim. Cosmochim. Acta*, **21**, 182.

Mackenzie, F. T. and Gees, R. (1971). *Science*, **173**, 533.

Moerman, (1938). As cited in Henricks, S. B. and Jefferson, M. E. (1938), op. cit.

Moore, T. C., Jr. (1969). *Geol. Soc. Amer. Bull.*, **80**, 2103.

Morey, G. W., Fournier, R. O., and Rowe, J. J. (1964). *J. Geophys. Res.*, **69**, 1995.

Muroya, M. and Kondo, S. (1970). *Bull. Chem. Soc. Japan*, **43**, 3453.

Nelson, D. M. (1975). "Uptake and Regeneration of Silicic Acid by Marine Phytoplankton". Unpublished PhD thesis, Univ. of Alaska, Fairbanks, Alaska, USA.

Nelson, D. M. and Goering, J. J. (1977a). *Deep-Sea Res.*, **24**, 65.

Nelson, D. M. and Goering, J. J. (1977b). *Analyt. Biochem.*, **78**, 139.

O'Connor, T. L. and Greenberg, S. A. (1958). *J. Phys. Chem.*, **62**, 1195.

Okamoto, G., Okura, T., and Goto, K. (1957). *Geochim. Cosmochim. Acta*, **10**, 123.

Patrick, (1951). As extensively referenced in Iler, R. (1955), op. cit.

Riemann, B. E. F., Lewin, J. C., and Volcani, B. E. F. (1966). *J. of Phycology*, **2**, 74.

Rieck, G. D. and Stevels, J. M. (1951). *J. Soc. Glass Tech.*, **35**, 285.

Segnit, E. R., Stevens, T. J. and Jones, J. B. (1965). *Geol. Soc. Australia J.*, **12(2)**, 211.

Siever, R. (1962). *J. Geol.*, **70**, 127.

Simpson, T. L. and Vaccaro, C. A. (1974). *J. Ultrastructural Res.*, **47**, 296.

Sosman, R. B. (1965). "The Phases of Silica". Rutgers Univ. Press, New Brunswick, NJ.

Stober, W. (1967). *In* "Equilibrium Concepts in Natural Water Systems". Am. Chem. Soc. Series 67, Washington DC.

Stoermer, E. F., Pankratz, H. S., and Bowen, C. C. (1965). *Am. J. Botany*, **52**, 1067.

Strelko, V. V., Burushkina, T. N., and Belyakov, V. N. (1974). *Doklady Akad. Nauk. S.S.S.R.*, **215(3)**, 606.

Taliaferro, N. L. (1935). *Am. J. Sci.*, **30**, 450.

Tanabe, K. (1970). "Solid Acids and Bases, Their Catalytic Properties". Academic Press, New York and London.

van Lier, J. A., de Bruyn, P. L. and Overbeek, J. Th. G. (1960). *J. Phys. Chem.*, **64**, 1675.

Verduch, A. G. (1958). *J. Am. Ceram. Soc.*, **41(11)**, 427.

Wade, W. (1964). *J. Phys. Chem.*, **68(5)**, 1029.

Wade, W. (1965). *J. Phys. Chem.*, **69(1)**, 322.

Weiler, R. and Mills, A. (1965). *Deep-Sea Res.*, **12**, 511.

Willey, J. (1974). *Mar. Chem.*, **2**, 239.

Willey, J. (1975a). *Mar. Chem.*, **3**, 227.

Willey, J. (1975b). *Mar. Chem.*, **3**, 241.

Wirth, and Gieskes, J. (1979). *J. Colloid and Interface Sci.*, **68(3)**, 492.

Wollast, R. (1974). *In* "The Sea", Vol. V—Marine Chemistry (Goldberg, E. D., ed.). Wiley-Interscience, New York.

Wollast, R. and Garrels, R. M. (1971). *Nature (Phys. Sci.)*, **229**, 94.

Zemmels, and Cook (1973). *In* "Initial Reports of the Deep Sea Drilling Project", Vol. 32 (Larson, R. L. *et al.*, eds.). US Government Printing Office, 547.

Zwietering, P. (1956). Proc. Inter. Symp. React. Solids, III.

Subject Index